Professional English for Material Science and Engineering
材料科学与工程专业英语

戴庆伟　杨青山　喻红梅　柴森森　周　涛　韩会全　编著

重庆大学出版社

内容提要

本书全面系统地讲述了金属材料及其加工制备的基础理论和加工工艺,内容涉及金属材料、热处理原理和工艺、塑性变形理论,以及铸造、挤压、轧制、锻压、焊接、模具成形等塑性成型的多种工艺。

本书最大的特点是阅读与翻译结合,阅读有词汇和问题,翻译有方法指导。本书分为两大部分,即专业英语阅读和专业英语翻译。每节阅读后有名词和短语中英文对照,便于学生掌握关键专业词汇。每节后面的问题便于检查学生的阅读理解情况。在第二部分还专门介绍了专业英语的特点和翻译方法,便于学生熟练阅读和理解专业文献。

本书可作为普通高等学校材料科学与工程、材料成型及控制工程、材料加工工程、金属材料工程等专业本科生和研究生的教材,也可作为从事材料相关行业人员进一步提高专业英语阅读能力的参考用书。

图书在版编目(CIP)数据

材料科学与工程专业英语 / 戴庆伟等编著. -- 重庆:
重庆大学出版社,2023.1
材料科学与工程专业本科系列教材
ISBN 978-7-5689-3329-2

Ⅰ.①材… Ⅱ.①戴… Ⅲ.①材料科学—英语—高等
学校—教材 Ⅳ.①TB3

中国版本图书馆 CIP 数据核字(2022)第 089215 号

材料科学与工程专业英语
CAILIAO KEXUE YU GONGCHENG ZHUANYE YINGYU

戴庆伟 杨青山 喻红梅 柴淼淼 周 涛 韩会全 编著
策划编辑:范 琪
特约编辑:薛婧媛

责任编辑:陈 力 版式设计:范 琪
责任校对:谢 芳 责任印制:张 策
*
重庆大学出版社出版发行
出版人:饶帮华
社址:重庆市沙坪坝区大学城西路 21 号
邮编:401331
电话:(023) 88617190 88617185(中小学)
传真:(023) 88617186 88617166
网址:http://www.cqup.com.cn
邮箱:fxk@ cqup.com.cn(营销中心)
全国新华书店经销
重庆市国丰印务有限责任公司印刷
*
开本:787mm×1092mm 1/16 印张:17 字数:527 千
2023 年 1 月第 1 版 2023 年 1 月第 1 次印刷
印数:1—2 000
ISBN 978-7-5689-3329-2 定价:49.80 元

前 言

　　为了满足材料科学与工程、材料成型及控制工程、材料加工工程、金属材料工程等专业学生学习需要,培养学生阅读和翻译专业文献的能力而编著本教材。教材中的英文原文段落多摘自英文原著,或来源于专门的英文学术网站,以让学生学习原汁原味的专业英语。

　　本教材最大的特点是阅读与翻译结合,阅读有词汇和问题,翻译有方法指导。专业英语阅读部分每节课后有专业词汇和术语,中英文对照,能让学生尽快掌握专业词汇和术语。每节后面的问题,能检查学生对阅读内容的理解和掌握情况。专业英语翻译部分讲述了专业英语的特点,英汉对比,以及专业英语翻译的方法,便于学生熟练阅读和理解专业文献。本书分为两个部分:专业英语阅读和专业英语翻译,共 37 个章节。内容涵盖金属材料、热处理原理、塑性成形原理,以及铸造、挤压、轧制、锻压、焊接、模具成形等多种工艺。本书设置章节多,内容广。各高校可根据各自学校人才培养的侧重点,选用合适的章节。

　　本书由重庆科技学院戴庆伟教授和重庆理工大学周涛教授组织协调编著。戴庆伟教授负责统稿,编写第 1、2、11、13 章。重庆科技学院杨青山副教授编写第 3—10、12、14—16 章,柴森森博士编写 17—35 章。成都工业学院喻红梅老师编写第二部分专业英语翻译。中冶赛迪研发中心有限公司的韩会全正高级工程师参与了第 13 章的编写和全书的编校。

　　由于编者水平和经验有限,书中如有错误和缺陷,恳请读者批评指正。

编著者

2022 年 1 月

目录

2

第二部分　专业英语翻译

第一部分 专业英语阅读

1

Metals and alloys

1.1 Metals

A metal is a material (an element, compound, or alloy) that is typically hard, opaque, shiny, and has good electrical and thermal conductivity. Metals are generally malleable—that is, they can be hammered or pressed permanently out of shape without breaking or cracking—as well as fusible (able to be fused or melted) and ductile (able to be drawn out into a thin wire). About 91 of the 118 elements in the periodic table are metals, and the others are nonmetals or metalloids. Some elements appear in both metallic and nonmetals forms.

Astrophysicists use the term "metal" to collectively describe all elements other than hydrogen and helium, the simplest two, in a star. The star fuses smaller atoms, mostly hydrogen and helium, to make larger ones over its lifetime. In that sense, the metallicity of an object is the proportion of its matter made up of all heavier chemical elements, not just traditional metals.

Many elements and compounds that are not normally classified as metals become metallic under high pressures; these are formed as metallic allotropes of nonmetals.

Words and terms

thermal conductivity　热传导性
malleable　有延展性的
fusible　可熔的
ductile　可延展的
metallic　金属的
allotropes　同素异形体

Questions

What is the property of metals?

How many metal elements in the world?

1.1.1 Structure and bonding

The atoms of metallic substances are typically arranged in one of three common crystal structures, namely body-centered cubic (BCC), face-centered cubic (FCC), and hexagonal close-packed (HCP)(Fig. 1.1). In BCC, each atom is positioned at the center of a cube of eight others. In FCC and HCP, each atom is surrounded by twelve others, but the stacking of the layers differs. Some metals adopt different structures depending on the temperature.

Atoms of metals readily lose their outer shell electrons, resulting in a free flowing cloud of electrons within their otherwise solid arrangement. This provides the ability of metallic substances to easily transmit heat and electricity. While this flow of electrons occurs, the solid characteristic of the metal is produced by electrostatic interactions between each atom and the electron cloud. This type of bond is called a metallic bond.

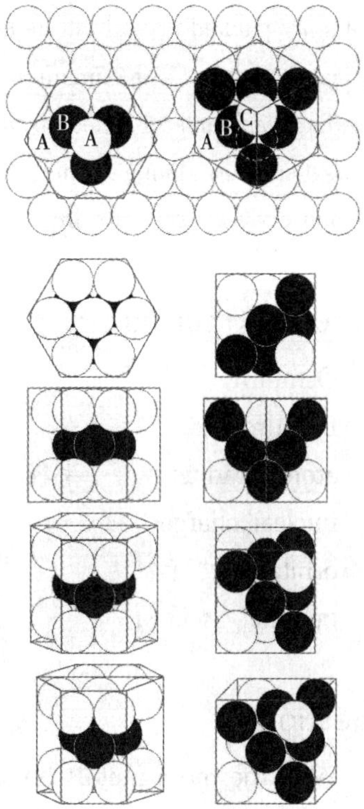

Fig. 1. 1 HCP and FCC close-packing of spheres

Words and terms

body-centered cubic (BCC) 体心立方

face-centered cubic (FCC) 面心立方

hexagonal close-packed (HCP) 密排六方

electrostatic interactions 静电相互作用

metallic bond 金属键

Questions

What are the typical crystal structures?

Why can the metals easily transmit heat and electricity?

1.1.2 Physical

Metals in general have high electrical conductivity, high thermal conductivity, and high density. Typically, they are malleable and ductile, deforming under stress without cleaving. In terms of optical properties, metals are shiny and lustrous. Sheets of metal beyond a few micrometers in thickness appear opaque, but gold leaf transmits green light.

Although most metals have higher densities than most nonmetals, there is wide variation in their densities, lithium being the least dense solid element and osmium the densest. The alkali and alkaline earth metals in groups ⅠA and ⅡA are referred to as the light metals because they have low density, low hardness, and low melting points. The high density of most metals is due to the tightly packed crystal lattice of the metallic structure. The strength of metallic bonds for different metals reaches a maximum around the center of the transition metal series, as those elements have large amounts of delocalized electrons in tight binding type metallic bonds. However, other factors (such as atomic radius, nuclear charge, number of bonds orbitals, overlap of orbital energies and crystal form) are involved as well.

Words and terms

Osmium　锇
Lithium　锂
atomic radius　原子半径
nuclear charge　核电荷
orbitals　电子轨道
opaque　不透明

Questions

Why the most metals have higher densities than most nonmetals?
What is the least dense solid element?

1.1.3　Mechanical

Mechanical properties of metals include ductility, i. e. their capacity for plastic deformation. Reversible elastic deformation in metals can be described by Hooke's Law for restoring forces, where the stress is linearly proportional to the strain. Forces larger than the elastic limit, or heat, may cause a permanent (irreversible) deformation of the object, known as plastic deformation or plasticity. This irreversible change in atomic arrangement may occur as a result of:

1) The action of an applied force (or work). An applied force may be tensile (pulling) force, compressive (pushing) force, shear, bending or torsion (twisting) forces.

2) A change in temperature. A temperature change may affect the mobility of the structural defects such as grain boundaries, point vacancies, line and screw dislocations, stacking faults and twins in both crystalline and non-crystalline solids. The movement or displacement of such mobile

defects is thermally activated, and thus limited by the rate of atomic diffusion.

3) Viscous flow near grain boundaries, for example, can give rise to internal slip, creep and fatigue in metals. It can also contribute to significant changes in the microstructure like grain growth and localized densification due to the elimination of intergranular porosity. Screw dislocations may slip in the direction of any lattice plane containing the dislocation, while the principal driving force for "dislocation climb" is the movement or diffusion of vacancies through a crystal lattice.

4) In addition, the nondirectional nature of metallic bonding is also thought to contribute significantly to the ductility of most metallic solids. When the planes of an ionic bond slide past one another, the resultant change in location shifts ions of the same charge into close proximity, resulting in the cleavage of the crystal; such shift is not observed in covalently bonded crystals where fracture and crystal fragmentation occurs.

Words and terms

mechanical properties　机械性能
plastic deformation　塑性变形
elastic deformation　弹性变形
creep　蠕变
fatigue　疲劳
intergranular　晶粒间的
porosity　多孔性
screw dislocations　螺型位错

Questions

What results to plastic deformation?
What can viscous flow effect?

1.2　Alloys

An alloy is a mixture of chemical elements, which forms an impure substance (admixture) that retains the characteristics of a metal. An alloy is distinct from an impure metal in that, with an alloy, the added elements are well controlled to produce desirable properties, while impure metals such as wrought iron, are less controlled, but are often considered useful. Alloys are made by mixing two or more elements, at least one of which is a metal. This is usually called the primary metal or the base metal, and the name of this metal may also be the name of the alloy. The other constituents may or may not be metals but, when mixed with the molten base, they will be soluble and dissolve into the mixture.

Words and terms

alloy 合金

chemical elements 化学元素

characteristics 性质

impure metal 含杂质金属

wrought iron 纯铁

base metal 基材

Questions

What is alloy?

What is the difference between impure metals and alloys?

1.2.1 Mechanical properties

The mechanical properties of alloy will often be quite different from those of its individual constituents. A metal that is normally very soft (malleable), such as aluminum, can be altered by alloying it with another soft metal, such as copper. Although both metals are very soft and ductile, the resulting aluminum alloy will have much greater strength. Adding a small amount of aluminum carbon to iron trades its great ductility for the greater strength of an alloy called steel. Due to its very-high strength, but still substantial toughness, and its ability to be greatly altered by heat treatment, steel is one of the most useful and common alloys in modern use. By adding chromium to steel, its resistance to corrosion can be enhanced, creating stainless steel, while adding silicon will alter its electrical characteristics, producing silicon steel.

Although the elements of an alloy usually must be soluble in the liquid state, they may not always be soluble in the solid state. If the metals remain soluble when solid, the alloy forms a solid solution, becoming a homogeneous structure consisting of identical crystals, called a phase. If the mixture cooling the constituents become insoluble, they may separate to form two or more different types of crystals, creating a heterogeneous microstructure of different phases, some with more of one constituent than the other phase has. However, in other alloys, the insoluble elements may not separate until crystallization occurs. If cooled very quickly, they first crystallize as a homogeneous phase, but they are supersaturated with the secondary constituents. As time passes, the atoms of these supersaturated alloys can separate from the crystal lattice, becoming more stable, and form a second phase that serve to reinforce the crystals internally.

Some alloys, such as electrum which is an alloy consisting of silver and gold, occur naturally. Meteorites are sometimes made of naturally occurring alloys of iron and nickel, but are not native to the Earth. One of the first alloys made by humans was bronze, which is a mixture of the metals tin and copper. Bronze was an extremely useful alloy to the ancients, because it is much stronger and harder than either of its components. Steel was another common alloy. However, in ancient times, it could only be created as an accidental byproduct from the heating of iron ore in fires (smelting) during the manufacture of iron. Other ancient alloys include pewter, brass and pig iron. In the modern age, steel can be created in many forms. Carbon steel can be made by varying only the carbon content, producing soft alloys like mild steel or hard alloys like spring steel. Alloy steels can be made by adding other elements, such as chromium, molybdenum, vanadium or nickel, resulting in alloys such as highspeed steel or tool steel. Small amounts of manganese are usually alloyed with most modern steels because of its ability to remove unwanted impurities, like phosphorus, sulfur and oxygen, which can have detrimental effects on the alloy. However, most alloys were not created until the 1,900 s, such as various aluminium, titanium, nickel, and magnesium alloys. Some modern superalloys, such as incoloy, inconel, and hastelloy, may consist of a multitude of different elements.

Words and terms

malleable　可塑的

aluminium　铝

copper　铜

ductile　延展性好的

heat treatment　热处理

chromium　铬

stainless steel　不锈钢

solid solution　固溶体

homogeneous structure　均一结构

crystal　晶体

phase　相

supersaturate　使……过饱和

electrum　琥珀金(金银合金)

meteorites　陨石

nickel　镍

bronze　青铜

byproduct　副产物

smelting　冶炼

pewter　白镴

brass　黄铜

pig iron　生铁

spring steel　弹簧钢

molybdenum　钼

vanadium 钒

phosphorus 磷

sulfur 硫黄

magnesium 镁

Questions

What are the characteristics of the mechanical properties of alloys?

What are solid solutions and phases?

How do alloy develop?

1.2.2　Alloying

Alloying a metal is done by combining it with one or more other elements that often enhance its properties. For example, the combination of carbon with iron produces steel, which is stronger than iron—its primary element. The electrical and thermal conductivity of alloys is usually lower than that of the pure metals. The physical properties, such as density, reactivity, and Young's modulus of an alloy may not differ greatly from those of its base element, but engineering properties such as tensile strength, ductility, and shear strength may be substantially different from those of the constituent materials. This is sometimes a result of sizes of atoms in the alloy, because larger atoms exert a compressive force on neighboring atoms, and smaller atoms exert a tensile force on their neighbors, helping the alloy resist deformation. Sometimes alloys may exhibit marked differences in behavior even when small amounts of one element are present. For example, impurities in semiconducting ferromagnetic alloys lead to different properties, as first predicted by White, Hogan, Suhl, Tian Abrie and Nakamura. Some alloys are made by melting and mixing two or more metals. Bronze, an alloy of copper and tin, was the first alloy discovered, during the prehistoric period now known as the bronze age. It was harder than pure copper and originally used to make tools and weapons, but was later superseded by metals and alloys with better properties. In later times bronze has been used for ornaments, bells, statues, and bearings. Brass is an alloy made from copper and zinc.

Unlike pure metals, most alloys do not have a single melting point, but a melting range during which the material is a mixture of solid and liquid phases (slush). The temperature at which melting begins is called the solidus, and the temperature when melting is just complete is called the liquidus. For many alloys there is a particular alloy proportion (in some cases more than one), called either a eutectic mixture or a peritectic composition, which gives the alloy a unique and low melting point, and no liquid/solid slush transition.

Words and terms

thermal conductivity 导热性

density　密度

reactivity　活性;反应性

Young's modulus　杨氏模量

tensile strength　抗拉强度

ductility　延展性

shear strength　剪切强度

ferromagnetic　铁磁的

solidus　固相线

liquidus　液相线

eutectic　共晶的

peritectic　包晶的

Questions

What are the advantages of alloy materials over pure metal materials?

What is the relationship between the properties of an alloy and its compositions?

1.2.3　Applications

Some metals and metal alloys possess high structural strength per unit mass, making them useful materials for carrying large loads or resisting impact damage. Metal alloys can be engineered to have high resistance to shear, torque and deformation. However, the same metal can also be vulnerable to fatigue damage through repeated use or from sudden stress failure when a load capacity is exceeded. The strength and resilience of metals has led to their frequent use in high-rise building and bridge construction, as well as most vehicles, many appliances, tools, pipes, non-illuminated signs and railroad tracks.

The two most commonly used structural metals, iron and aluminium, are also the most abundant metals in the Earth's crust.

Metals are good conductors, making them valuable in electrical appliances and for carrying an electric current over a distance with little energy lost. Electrical power grids rely on metal cables to distribute electricity. Home electrical systems, for the most part, are wired with copper wire for its good conducting properties.

The thermal conductivity of metal is useful for containers to heat materials over a flame. Metal is also used for heat sinks to protect sensitive equipment from overheating.

The high reflectivity of some metals is important in the construction of mirrors, including precision astronomical instruments. This last property can also make metallic jewelry aesthetically appealing.

Some metals have specialized uses; radioactive metals such as uranium and plutonium are used in nuclear power plants to produce energy via nuclear fission. Mercury is a liquid at room temperature and is used in switches to complete a circuit when it flows over the switch contacts. Shape memory alloy is used for applications such as pipes, fasteners and vascular stents.

Metals can be doped with foreign molecules—organic, inorganic, biological and polymers. This doping entails the metal with new properties that are induced by the guest molecules. Applications in catalysis, medicine, electrochemical cells, corrosion and more have been developed.

Words and terms

torque 扭转

vulnerable 脆弱的

fatigue damage 疲劳破坏

resilience 弹性

Earth's crust 地壳

electric current 电流

thermal conductivity 导热性

heat sinks 散热片

reflectivity 反射率

precision astronomical instruments 精密的天文仪器

aesthetically 审美地

radioactive metals 放射性金属

uranium 铀

plutonium 钚

nuclear fission 核裂变

mercury 水银

shape memory alloy 形状记忆合金

vascular stents 血管支架

dope 掺杂

catalysis 催化剂

Questions

Why can metal alloys be used as structural materials?

What are the characteristics of heat conducting material?

1.3　Ferrous metals and alloys

1.3.1　Steel

Steel is an alloy of iron and other elements, primarily carbon, that is widely used in construction and other applications because of its high tensile strength and low cost. Steel's base metal is iron, which is able to take on two crystalline forms (allotropic forms), body centered cubic (BCC) and face centered cubic (FCC), depending on its temperature. It is the interaction of those allotropes with the alloying elements, primarily carbon, that gives steel and cast iron their range of unique properties. In the body-centred cubic arrangement, there is an iron atom in the centre of each cube, and in the face-centred cubic, there is one at the center of each of the six faces of the cube. Carbon, other elements, and inclusions within iron act as hardening agents that prevent the movement of dislocations that otherwise occur in the crystal lattices of iron atoms.

The carbon in typical steel alloys may contribute up to 2.1% of its weight. Varying the amount of alloying elements, their presence in the steel either as solute elements, or as precipitated phases, retards the movement of those dislocations that make iron comparatively ductile and weak, and thus controls its qualities such as the hardness, ductility, and tensile strength of the resulting steel. Steel's strength compared to pure iron is only possible at the expense of iron's ductility, of which iron has an excess.

Steel was produced in bloomery furnaces for thousands of years, but its extensive use began after more efficient production methods were devised in the 17th century, with the production of blister steel and then crucible steel. With the invention of the Bessemer process in the mid-19th century, a new era of mass-produced steel began. This was followed by Siemens-Martin process and then Gilchrist-Thomas process that refined the quality of steel. With their introductions, mild steel replaced wrought iron.

Further refinements in the process, such as basic oxygen steelmaking (BOS), largely replaced earlier methods by further lowering the cost of production and increasing the quality of the product. Today, steel is one of the most common materials in the world, with more than 1.3×10^9 t produced annually. It is a major component in buildings, infrastructure, tools, ships, automobiles, machines, appliances, and weapons. Modern steel is generally identified by various grades defined by assorted standards organizations.

The carbon content of steel is between 0.002% and 2.1% by weight for plain iron-carbon alloys. These values vary depending on alloying elements such as manganese, chromium, nickel, iron, tungsten, carbon and so on. Basically, steel is an iron-carbon alloy that does not undergo eutectic reaction. In contrast, cast iron does undergo eutectic reaction. Too little carbon content leaves (pure) iron quite soft, ductile, and weak. Carbon contents higher than those of steel make an alloy, commonly called pig iron, that is brittle (not malleable). While iron alloyed with carbon is called carbon steel, alloy steel is steel to which other alloying elements have been intentionally added to modify the characteristics of steel. Common alloying elements include: manganese, nickel, chromium, molybdenum, boron, titanium, vanadium, tungsten, cobalt, and niobium. Additional elements are also important in steel: phosphorus, sulfur, silicon, and traces of oxygen, nitrogen, and copper, which are most frequently considered undesirable.

Alloys with a higher than 2.1% carbon content, depending on other element content and possibly on processing, are known as cast iron. Cast iron is not malleable even when hot, but it can be formed by casting as it has a lower melting point than steel and good castability properties. Certain compositions of cast iron, while retaining the economies of melting and casting, can be heat treated after casting to make malleable iron or ductile iron objects. Steel is also distinguishable from wrought iron (now largely obsolete), which may contain a small amount of carbon but large amounts of slag.

Iron is commonly found in the Earth's crust in the form of an ore, usually an iron oxide, such as magnetite, hematite, etc. Iron is extracted from iron ore by removing the oxygen through its combination with a preferred chemical partner such as carbon that is then lost to the atmosphere as carbon dioxide. This process, known as smelting, was first applied to metals with lower melting points, such as tin, which melts at about 250 ℃, and copper, which melts at about 1,100 ℃ and the combination, bronze, which is liquid at less than 1,083 ℃. In comparison, cast iron melts at about 1,375 ℃. Small quantities of iron were smelted in ancient times, in the solid state, by heating the ore in a charcoal fire and then welding the clumps together with a hammer and in the process squeezing out the impurities. With care, the carbon content could be controlled by moving it around in the fire. Unlike copper and tin, liquid or solid iron dissolves carbon quite readily.

All of these temperatures could be reached with ancient methods used since the Bronze Age. Since the oxidation rate of iron increases rapidly beyond 800 ℃, it is important that smelting take place in a low-oxygen environment. Smelting, using carbon to reduce iron oxides, results in an alloy (pig iron) that retains too much carbon to be called steel. The excess carbon and other impurities are removed in a subsequent step.

Other materials are often added to the iron/carbon mixture to produce steel with desired properties. Nickel and manganese in steel add to its tensile strength and make the austenite form of the iron-carbon solution more stable, chromium increases hardness and melting temperature, and vanadium also increases hardness while making it less prone to metal fatigue.

To inhibit corrosion, at least 11% chromium is added to steel so that a hard oxide forms on the metal surface, which is known as stainless steel. Tungsten slows the formation of cementite, keeping carbon in the iron matrix and allowing martensite to preferentially form at slower quench rates, resulting in high speed steel. On the other hand, sulfur, nitrogen, and phosphorus are considered contaminants that make steel more brittle and are removed from the steel melt during processing.

Even in a narrow range of concentrations of mixtures of carbon and iron that make a steel, a number of different metallurgical structures, with very different properties can form. Understanding such properties is essential to making quality steel. At room temperature, the most stable form of pure iron is the body-centered cubic (BCC) structure called alpha iron or α-iron. It is a fairly soft metal that can dissolve only a small concentration of carbon, no more than 0.005% at 0 ℃ and 0.020% in terms of mass fraction at 723 ℃. The inclusion of carbon in alpha iron is called ferrite. At 911 ℃ pure iron transforms into a face-centered cubic (FCC) structure, called gamma iron or γ-iron. The inclusion of carbon in gamma iron is called austenite. The more open FCC structure of austenite, considerably more carbon can be dissolved, as much as 2.06% (38 times that of ferrite) carbon at 1,147 ℃, which reflects the upper carbon content of steel, beyond which is cast iron. When carbon moves out of solution with iron it forms a very hard, but brittle material called cementite (Fe_3C).

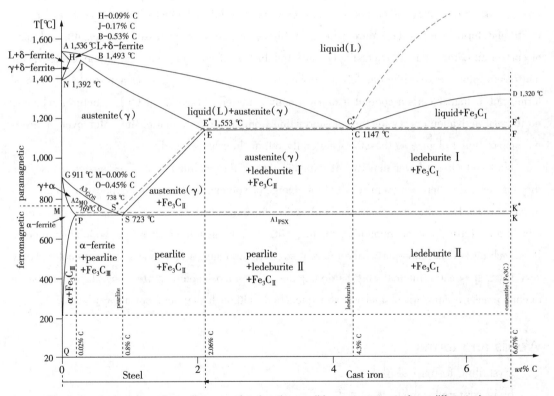

Fig. 1.2   Iron-carbon phase diagram, showing the conditions necessary to form different phases

When steels with exactly 0.8% carbon (known as a eutectoid steel) are cooled, the austenitic phase (FCC) of the mixture attempts to revert to the ferrite phase (BCC). The carbon no longer fits within the FCC austenite structure, resulting in an excess of carbon. One way for carbon to leave the austenite is for it to precipitate out of solution as cementite, leaving behind a surrounding phase of BCC iron called ferrite with a small percentage of carbon in solution. The two, ferrite and cementite, precipitate simultaneously producing a layered structure called pearlite, named for its resemblance to mother of pearl. In a hypereutectoid composition (greater than 0.8% carbon), the carbon will first precipitate out as large inclusions of cementite at the austenite grain boundaries until the percentage of carbon in the grains has decreased to the eutectoid composition (0.8% carbon), at which point the pearlite structure forms. For steels that have less than 0.8% carbon (hypoeutectoid), ferrite will first form within the grains until the remaining composition rises to 0.8% of carbon, at which point the pearlite structure will form. No large inclusions of cementite will form at the boundaries in hypoeutectoid steel. The above assumes that the cooling process is very slow, allowing enough time for the carbon to migrate.

As the rate of cooling is increased, the carbon will have less time to migrate to form carbide at the grain boundaries but will have increasingly large amounts of pearlite of a finer and finer structure within the grains; hence the carbide is more widely dispersed and acts to prevent slip of defects within those grains, resulting in hardening of the steel. At the very high cooling rates produced by quenching, the carbon has no time to migrate but is locked within the face center austenite and forms martensite. Martensite is a highly strained and stressed, supersaturated form of carbon and iron, and is exceedingly hard but brittle. Depending on the carbon content, the martensitic phase takes different forms. Below 0.2% carbon, it takes on a ferrite BCC crystal form, but at higher carbon content it takes a body-centered tetragonal (BCT) structure. There is no thermal activation energy for the transformation from austenite to martensite. Moreover, there is no compositional change so the atoms generally retain their same neighbors.

Martensite has a lower density (it expands during the cooling) than does austenite, so that the transformation between them results in a change of volume. In this case, expansion occurs. Internal stresses from this expansion generally take the form of compression on the crystals of martensite and tension on the remaining ferrite, with a fair amount of shear on both constituents. If quenching is done improperly, the internal stresses can cause a part to shatter as it cools. At the very least, they cause internal work hardening and other microscopic imperfections. It is common for quench cracks to form when steel is water quenched, although they may not always be visible.

Words and terms

crystalline forms   晶型
body centered cubic   体心立方
face centered cubic   面心立方
allotrope   同素异形体
inclusion   杂质
lattice   晶格
atom   原子

precipitated phase 析出相

retard 妨碍

bloomery 土法熟铁吹炼炉

blister steel 泡钢（由熟铁渗碳而成的钢）

crucible 坩埚

infrastructure 基础建设

tungsten 钨

boron 硼

titanium 钛

cobalt 钴

niobium 铌

magnetite 磁铁矿

hematite 赤铁矿

charcoal 木炭

austenite 奥氏体

cementite 渗碳体

quench 淬火

metallurgical 金相的

martensite 马氏体

ferrite 铁素体

pearlite 珠光体

hypereutectoid 过共析的

hypoeutectoid 亚共析的

carbide 碳化物

Questions

What is the difference between steel and cast iron?

How does pearlite form in the cooling process of eutectoid steel?

1.3.2 Carbon steels

Carbon steels are steels with carbon content up to 2.1% by weight. American Iron and Steel Institute (AISI) definition of Carbon Steel states:

Steel is considered to be carbon steel when no minimum content is specified or required for chromium, cobalt, molybdenum, nickel, niobium, titanium, tungsten, vanadium or zirconium, or any other element to be added to obtain a desired alloying effect;

The specified minimum for copper does not exceed 0.40%; or the maximum content specified for any of the following elements does not exceed the percentages noted: manganese 1.65, silicon

0.60, copper 0.60.

The term "carbon steel" may also be used in reference to steel which is not stainless steel; in this use carbon steel may include alloy steels.

As the carbon percentage content rises, steel has the ability to become harder and stronger through heat treating; however, it becomes less ductile. Regardless of the heat treatment, a higher carbon content reduces weldability. In carbon steels, the higher carbon content lowers the melting point.

(1) Mild or low-carbon steel

Mild steel (steel containing a small percentage of carbon, strong and tough but not readily tempered), also known as plain-carbon steel and low carbon steel. It is now the most common form of steel because its price is relatively low while it provides material properties that are acceptable for many applications. Mild steel contains approximately 0.05% to 0.25% carbon making it malleable and ductile. Mild steel has a relatively low tensile strength, but it is cheap and easy to form; surface hardness can be increased through carburizing.

It is often used when large quantities of steel are needed, for example as structural steel. The density of mild steel is approximately 7.85 g/cm^3 (7,850 kg/m^3 or 0.284 lb/in^3) and the Young's modulus is 200 GPa.

Low-carbon steels suffer from yield-point runout where the material has two yield points. The first yield point (or upper yield point) is higher than the second and the yield drops dramatically after the upper yield point. If a low-carbon steel is only stressed to some point between the upper and lower yield point then the surface develops Lüder bands. Low-carbon steels contain less carbon than other steels and are easier to cold-form, making them easier to handle.

(2) Higher-carbon steels

Carbon steels which can successfully undergo heat-treatment have a carbon content in the range of 0.30% to 1.70% by weight. Trace impurities of various other elements can have a significant effect on the quality of the resulting steel. Trace amounts of sulfur in particular make the steel red-short, that is, brittle and crumbly at working temperatures. Low-alloy carbon steel, such as A36 grade, contains about 0.05% sulfur and melts around 1,426 ℃ to 1,538 ℃ (2,599 ℉ to 2,800 ℉). Manganese is often added to improve the hardenability of low-carbon steels. These additions turn the material into a low-alloy steel by some definitions, but AISI's definition of carbon steel allows up to 1.65% manganese by weight.

Words and terms

mild steel　软钢

weldability　可焊性

temper　回火

plain-carbon steel　普通碳素钢

carburizing 渗碳
yield-point 屈服点
red-short 热脆的

Questions

What is the definition of carbon steel?

What's the difference between low-carbon steel and high-carbon steel?

1.3.3 Alloy steels

Alloy steel is steel that is alloyed with a variety of elements in total amounts between 1.0% and 50% by weight to improve its mechanical properties. Alloy steels are broken down into two groups: low-alloy steels and high-alloy steels. The difference between the two is somewhat arbitrary: Smith and Hashemi define the difference at 4.0%, while Degarmo et al., define it at 8.0%. Most commonly, the phrase "alloy steel" refers to low-alloy steels.

Strictly speaking, every steel is alloy, but not all steels are called "alloy steels". The simplest steels are iron (Fe) alloyed with carbon (C) (about 0.1% to 1.0%, depending on type). However, the term "alloy steel" is the standard term referring to steels with other alloying elements added deliberately in addition to the carbon. Common alloyants include manganese (the most common one), nickel, chromium, molybdenum, vanadium, silicon, and boron. Less common alloyants include aluminum, cobalt, copper, cerium, niobium, titanium, tungsten, tin, zinc, lead, and zirconium.

The following is a range of improved properties in alloy steels (as compared to carbon steels): strength, hardness, toughness, wear resistance, corrosion resistance, hardenability, and hot hardness. To achieve some of these improved properties the metal may require heat treating.

Some of these find uses in exotic and highly-demanding applications, such as in the turbine blades of jet engines, in spacecraft, and in nuclear reactors. Because of the ferromagnetic properties of iron, some steel alloys find important applications where their responses to magnetism are very important, including in electric motors and in transformers.

High-strength low-alloy steel (HSLA) is a type of alloy steel that provides better mechanical properties or greater resistance to corrosion than carbon steel. HSLA steels vary from other steels in that they are not made to meet a specific chemical composition but rather to specific mechanical properties. They have a carbon content between 0.05% to 0.25% to retain formability and weldability. Other alloying elements include up to 2.0% manganese and small quantities of copper, nickel, niobium, nitrogen, vanadium, chromium, molybdenum, titanium, calcium, rare earth elements, or zirconium. Copper, titanium, vanadium, and niobium are added for strengthening purposes. These elements are intended to alter the microstructure of carbon steels, which is usually a ferrite-pearlite aggregate, to produce a very fine dispersion of alloy carbides in an almost pure ferrite matrix. This eliminates the toughness-reducing effect of a pearlitic volume fraction yet maintains and increases the material's strength by refining the grain size, which in the case of ferrite increases yield strength by 50% for every halving of the mean grain diameter. Precipitation strengthening plays a minor role, too. Their yield strengths can be anywhere between 250 to 590 megapascals. Because of their higher strength and toughness HSLA steels usually require 25% to 30% more power to form, as compared to carbon steels.

Copper, silicon, nickel, chromium, and phosphorus are added to increase corrosion resistance. Zirconium, calcium, and rare earth elements are added for sulfide-inclusion shape control which increases formability. These are needed because most HSLA steels have directionally sensitive properties. Formability and impact strength can vary significantly when tested longitudinally and transversely to the grain. Bends that are parallel to the longitudinal grain are more likely to crack around the outer edge because it experiences tensile loads. This directional characteristic is substantially reduced in HSLA steels that have been treated for sulfide shape control.

They are used in cars, trucks, cranes, bridges, roller coasters and other structures that are designed to handle large amounts of stress or need a good strength-to-weight ratio. HSLA steel cross-sections and structures are usually 20% to 30% lighter than a carbon steel with the same strength.

HSLA steels are also more resistant to rust than most carbon steels because of their lack of pearlite - the fine layers of ferrite (almost pure iron) and cementite in pearlite. HSLA steels usually have densities of around $7,800 \text{ kg/m}^3$.

Words and terms

cerium   铈

exotic   外来的

turbine blade   涡轮机叶片

jet engine   喷气式发动机

nuclear reactor   核反应堆

electric motor   电动机

nitrogen   氮

calcium   钙

megapascal   兆帕

sulfide-inclusion   硫化夹杂物

longitudinally   纵向地

transversely   横向地

truck   卡车

crane   起重机

strength-to-weight ratio   比强度

Questions

Why is every steel considered as alloy?

Why does the reduction of pearlite lead to an increase in the corrosion resistance of steel?

1.3.4 Cast iron

Cast iron is a group of iron-carbon alloys with a carbon content greater than 2%. Its usefulness derives from its relatively low melting temperature. The alloy constituents affect its colour when fractured: white cast iron has carbide impurities which allow cracks to pass straight through, grey cast iron has graphite flakes which deflect a passing crack and initiate countless new cracks as the material breaks, and ductile cast iron has spherical graphite "nodules" which stop the crack from further progressing.

Carbon (C) ranging from 1.8 $wt\%$ to 4.0 $wt\%$, and silicon (Si) 1 $wt\%$ to 3 $wt\%$ are the main alloying elements of cast iron. Iron alloys with less carbon content are known as steel. While this technically makes the Fe-C-Si system ternary, the principle of cast iron solidification can be understood from the simpler binary iron-carbon phase diagram. Since the compositions of most cast irons are around the eutectic point (lowest liquid point) of the iron-carbon system, the melting temperatures usually range from 1,150 ℃ to 1,200 ℃, which is about 300 ℃ lower than the melting point of pure iron.

Cast iron tends to be brittle, except for malleable cast irons. With its relatively low melting point, good fluidity, castability, excellent machinability, resistance to deformation and wear resistance, cast irons have become an engineering material with a wide range of applications and are used in pipes, machines and automotive industry parts, such as cylinder heads (declining usage), cylinder blocks and gearbox cases (declining usage). It is resistant to destruction and weakening by oxidation.

The earliest cast iron artefacts date to the 5th century BC, and were discovered by archaeologists in what is now Jiangsu in China. Cast iron was used in ancient China for warfare, agriculture, and architecture. During the 15th century, cast iron became utilized for artillery in Burgundy, France, and in England during the Reformation. The first cast iron bridge was built during the 1770s by Abraham Darby Ⅲ, and is known as The Iron Bridge. Cast iron is also used in the construction of buildings.

Cast iron's properties are changed by adding various alloying elements, or alloyants. Next to carbon, silicon is the most important alloyant because it forces carbon out of solution. A low percentage of silicon allows carbon to remain in solution forming iron carbide and the production of white cast iron. A high percentage of silicon forces carbon out of solution forming graphite and the production of grey cast iron. Other alloying agents, manganese, chromium, molybdenum, titanium and vanadium counteracts silicon, promotes the retention of carbon, and the formation of those carbides. Nickel and copper increase strength, and machinability, but do not change the amount of graphite formed. The carbon in the form of graphite results in a softer iron, reduces shrinkage, lowers strength, and decreases density. Sulfur, largely a contaminant when present, forms iron sulfide, which prevents the formation of graphite and increases hardness. The problem with sulfur is that it makes molten cast iron viscous, which causes defects. To counter the effects of sulfur, manganese is added because the two form into manganese sulfide instead of iron sulfide. The manganese sulfide is lighter than the melt so it tends to float out of the melt and into the slag. The amount of manganese required to neutralize sulfur is 1.7×sulfur content+0.3%. If more than this amount of manganese is added, then manganese carbide forms, which increases hardness and chilling, except in grey iron, where up to 1% of manganese increases strength and density.

Nickel is one of the most common alloying elements because it refines the pearlite and graphite structure, improves toughness, and evens out hardness differences between section thicknesses. Chromium is added in small amounts to reduce free graphite and produce chill, because it is a powerful carbide stabilizer; nickel is often added in conjunction. A small amount of tin can be added as a substitute for 0.5% chromium. Copper is added in the ladle or in the furnace, on the order of 0.5% to 2.5%, to decrease chill, refine graphite, and increase fluidity. Molybdenum is added on the order of 0.3% to 1.0% to increase chill and refine the graphite and pearlite structure; it is often added in conjunction with nickel, copper, and chromium to form high strength irons. Titanium is added as a degasser and deoxidizer, but it also increases fluidity. 0.15% to 0.5% vanadium is added to cast iron to stabilize cementite, increase hardness, and increase resistance to wear and heat. 0.1% to 0.3% zirconium helps to form graphite, deoxidize, and increase fluidity.

In malleable iron melts, bismuth is added, on the scale of 0.002% to 0.010%, to increase how much silicon can be added. In white iron, boron is added to aid in the production of malleable iron; it also reduces the coarsening effect of bismuth.

(1) Grey cast iron

Grey cast iron is characterized by its graphitic microstructure, which causes fractures of the material to have a grey appearance. It is the most commonly used cast iron and the most widely used cast material based on weight. Most cast irons have a chemical composition of 2.5% to 4.0% carbon, 1% to 3% silicon, and the remainder iron. Grey cast iron has less tensile strength and shock resistance than steel, but its compressive strength is comparable to low- and medium-carbon steel. These mechanical properties are controlled by the size and shape of the graphite flakes present in the microstructure and can be characterized according to the guidelines given by the ASTM.

(2) White cast iron

White cast iron displays white fractured surfaces due to the presence of an iron carbide precipitate called cementite. With a lower silicon content (graphitizing agent) and faster cooling rate, the carbon in white cast iron precipitates out of the melt as the metastable phase cementite, Fe_3C, rather than graphite. The cementite which precipitates from the melt forms as relatively large particles. As the iron carbide precipitates out, it withdraws carbon from the original melt, moving the mixture toward one that is closer to eutectic, and the remaining phase is the lower iron-carbon austenite (which on cooling might transform to martensite). These eutectic carbides are much too large to provide the benefit of what is called precipitation hardening (as in some steels, where much smaller cementite precipitates might inhibit plastic deformation by impeding the movement of dislocations through the pure iron ferrite matrix). Rather, they increase the bulk hardness of the cast iron simply by virtue of their own very high hardness and their substantial volume fraction, such that the bulk hardness can be approximated by a rule of mixtures. In any case, they offer hardness at the expense of toughness. Since carbide makes up a large fraction of the material, white cast iron could reasonably be classified as a cermet. White iron is too brittle for use in many structural components, but with good hardness and abrasion resistance and relatively low cost, it finds use in such applications as the wear surfaces (impeller and volute) of slurry pumps, shell liners and lifter bars in ball mills and autogenous grinding mills, balls and rings in coal pulverisers, and the teeth of a backhoe's digging bucket (although cast medium-carbon martensitic steel is more common for this application).

It is difficult to cool thick castings fast enough to solidify the melt as white cast iron all the way through. However, rapid cooling can be used to solidify a shell of white cast iron, after which the remainder cools more slowly to form a core of grey cast iron. The resulting casting, called a chilled casting, has the benefits of a hard surface with a somewhat tougher interior.

High-chromium white iron alloys allow massive castings (for example, a 10-tonne impeller) to be sand cast, as the chromium reduces cooling rate required to produce carbides through the greater thicknesses of material. Chromium also produces carbides with impressive abrasion resistance. These high-chromium alloys attribute their superior hardness to the presence of chromium carbides. The main forms of these carbides are the eutectic or primary M_7C_3 carbides, where "M" represents iron or chromium and can vary depending on the alloy's composition. The eutectic carbides form as bundles of hollow hexagonal rods and grow perpendicular to the hexagonal basal plane. The hardness of these carbides is within the range of 1,500 HV to 1,800 HV.

(3) Malleable cast iron

Malleable iron starts as a white iron casting that is then heat treated for a day or two at about 950 ℃ and then cooled over a day or two. As a result, the carbon in iron carbide transforms into graphite and ferrite plus carbon (austenite). The slow process allows the surface tension to form the graphite into spheroidal particles rather than flakes. Due to their lower aspect ratio, the spheroids are relatively short and far from one another, and have a lower cross section vis-a-vis a propagating crack or phonon. They also have blunt boundaries, as opposed to flakes, which alleviates the stress concentration problems found in grey cast iron. In general, the properties of malleable cast iron are more like those of mild steel. There is a limit to how large a part can be cast in malleable iron, as it is made from white cast iron.

(4) Ductile cast iron

Developed in 1948, nodular or ductile cast iron has its graphite in the form of very tiny nodules with the graphite in the form of concentric layers forming the nodules. As a result, the properties of ductile cast iron are that of a spongy steel without the stress concentration effects that flakes of graphite would produce. Tiny amounts of 0.02% to 0.10% magnesium, and only 0.02% to 0.04% cerium added to these alloys slow the growth of graphite precipitates by bonding to the edges of the graphite planes. Along with careful control of other elements and timing, this allows the carbon to separate as spheroidal particles as the material solidifies. The properties are similar to malleable iron, but parts can be cast with larger sections.

Words and terms

graphite   石墨

flake   薄片

spherical   球状的

ternary   三元的

gearbox   变速箱

archaeologist   考古学家

artillery   火炮

shrinkage   收缩率

viscous   黏稠的

slag   矿渣

neutralize   中和

chill   激冷

degasser   脱气剂

bismuth   铋

cermet   金属陶瓷

abrasion   磨损

impeller   叶轮

volute   泵壳

slurry   浆体

liner   衬里

ball mill   球磨机

pulveriser   磨粉机

backhoe   反铲挖土机

bucket   铲斗

Questions

Why are the fracture colors of white cast iron and gray iron different?

What is the influence of the form of carbon on the properties of cast iron?

1.4 Non-ferrous metals and alloys

In metallurgy, a non-ferrous metal is a metal, including alloys, that does not contain iron (ferrite) in appreciable amounts. Generally, more expensive than ferrous metals, non-ferrous metals are used because of desirable properties such as low weight (e. g. aluminium), higher conductivity (e. g. copper), non-magnetic property or resistance to corrosion (e. g. zinc). Some non-ferrous materials are also used in the iron and steel industries. For example, bauxite is used as flux for blast furnaces, while others such as wolframite, pyrolusite and chromite are used in making ferrous alloys.

Important non-ferrous metals include aluminium, copper, lead, nickel, tin, titanium and zinc, and alloys such as brass. Precious metals such as gold, silver and platinum and exotic or rare metals such as cobalt, mercury, tungsten, beryllium, bismuth, cerium, cadmium, niobium, indium, gallium, germanium, lithium, selenium, tantalum, tellurium, vanadium, and zirconium are also non-ferrous. They are usually obtained through minerals such as sulfides, carbonates, and silicates. Non-ferrous metals are usually refined through electrolysis.

Words and terms

non-ferrous metal 有色金属材料

metallurgy 冶金学

bauxite 铝土矿

wolframite 黑钨矿

pyrolusite 软锰矿

chromite 铬铁矿

platinum 铂金

cadmium 镉

indium 铟

gallium 镓

germanium 锗

lithium 锂

selenium 硒

tantalum 钽

tellurium 碲

mineral 矿石

Questions

What are the common non-ferrous metals you know?

How is aluminum produced in the factory?

1.4.1 Aluminium

Aluminium or aluminum (in North American English) is a chemical element in the boron group with symbol Al and atomic number 13. It is a silvery-white, soft, nonmagnetic, ductile metal. By mass, aluminium makes up about 8% of the Earth's crust; it is the third most abundant element after oxygen and silicon and the most abundant metal in the crust, though it is less common in the mantle below. Aluminium metal is so chemically reactive that native specimens are rare and limited to extreme reducing environments. Instead, it is found combined in over 270 different minerals. The chief ore of aluminium is bauxite.

Aluminium is remarkable for the metal's low density and its ability to resist corrosion through the phenomenon of passivation. Aluminium and its alloys are vital to the aerospace industry and important in transportation and structures, such as building facades and window frames. The oxides and sulfates are the most useful compounds of aluminium.

Despite its prevalence in the environment, no known form of life uses aluminium salts metabolically, but aluminium is well tolerated by plants and animals. Because of these salts' abundance, the potential for a biological role for them is of continuing interest, and studies continue.

Aluminium is a relatively soft, durable, lightweight, ductile, and malleable metal with appearance ranging from silvery to dull gray, depending on the surface roughness. It is nonmagnetic and does not easily ignite. A fresh film of aluminium serves as a good reflector (approximately 92%) of visible light and an excellent reflector (as much as 98%) of medium and far infrared radiation. The yield strength of pure aluminium is 7 to 11 MPa, while aluminium alloys have yield strengths ranging from 200 MPa to 600 MPa. Aluminium has about one-third the density and stiffness of steel. It is easily machined, cast, drawn and extruded.

Aluminium atoms are arranged in a face-centered cubic (FCC) structure. Aluminium has a stacking-fault energy of approximately 200 mJ/m^2.

Aluminium is a good thermal and electrical conductor, having 59% the conductivity of copper, both thermal and electrical, while having only 30% of copper's density. Aluminium is capable of superconductivity, with a superconducting critical temperature of 1.2 kelvin and a critical magnetic field of about 100 gauss (10 milliteslas). Aluminium is the most common material for the fabrication of superconducting qubits.

Words and terms

mantle　地幔

passivation　钝化

metabolically　代谢地

ignite　点燃

infrared　红外的

stiffness　刚度

stacking-fault　堆垛层错

gauss　高斯

milliteslas　毫特斯拉

Questions

What are the applications of aluminum?

What are the characteristics of aluminum compared to steel?

1.4.2　Copper

Copper is a chemical element with symbol Cu and atomic number 29. It is a soft, malleable, and ductile metal with very high thermal and electrical conductivity. A freshly exposed surface of pure copper has a reddish-orange color. Copper is used as a conductor of heat and electricity, as a building material, and as a constituent of various metal alloys, such as sterling silver used in jewelry, cupronickel used to make marine hardware and coins, and constantan used in strain gauges and thermocouples for temperature measurement.

Copper is one of the few metals that occur in nature in directly usable metallic form as opposed to needing extraction from an ore. This led to very early human use, from 8,000 BC. It was the first metal to be smelted from its ore, 5,000 BC, the first metal to be cast into a shape in a mold, 4,000 BC and the first metal to be purposefully alloyed with another metal, tin, to create bronze, 3,500 BC.

The commonly encountered compounds are copper(Ⅱ) salts, which often impart blue or green colors to such minerals as azurite, malachite, and turquoise, and have been used widely and historically as pigments. Copper used in buildings, usually for roofing, oxidizes to form a green verdigris (or patina). Copper is sometimes used in decorative art, both in its elemental metal form and in compounds as pigments. Copper compounds are used as bacteriostatic agents, fungicides, and wood preservatives.

Copper is essential to all living organisms as a trace dietary mineral because it is a key constituent of the respiratory enzyme complex cytochrome C oxidase. In molluscs and crustaceans, copper is a constituent of the blood pigment hemocyanin, replaced by the iron-complexed hemoglobin in fish and other vertebrates. In humans, copper is found mainly in the liver, muscle, and bone. The adult body contains between 1.4 mg and 2.1 mg of copper per kilogram of body weight.

Copper, silver, and gold are in group 11 of the periodic table; these three metals have one s-orbital electron on top of a filled d-electron shell and are characterized by high ductility, and electrical and thermal conductivity. The filled d-shells in these elements contribute little to inter-atomic interactions, which are dominated by the s-electrons through metallic bonds. Unlike metals with incomplete d-shells, metallic bonds in copper are lacking a covalent character and are relatively weak. This observation explains the low hardness and high ductility of single crystals of copper. At the macroscopic scale, introduction of extended defects to the crystal lattice, such as grain boundaries, hinders flow of the material under applied stress, thereby increasing its hardness. For this reason, copper is usually supplied in a fine-grained polycrystalline form, which has greater strength than monocrystalline forms.

The softness of copper partly explains its high electrical conductivity (59.6×106 S/m) and high thermal conductivity, second highest (second only to silver) among pure metals at room temperature. This is because the resistivity to electron transport in metals at room temperature originates primarily from scattering of electrons on thermal vibrations of the lattice, which are relatively weak in a soft metal. The maximum permissible current density of copper in open air is approximately 3.1×106 A/m^2 of cross-sectional area, above which it begins to heat excessively.

Copper is one of a few metallic elements with a natural color other than gray or silver. Pure copper is orange-red and acquires a reddish tarnish when exposed to air. The characteristic color of copper results from the electronic transitions between the filled 3d and half-empty 4s atomic shells - the energy difference between these shells corresponds to orange light. As with other metals, if copper is put in contact with another metal, galvanic corrosion will occur.

Most copper is mined or extracted as copper sulfides from large open pit mines in porphyry copper deposits that contain 0.4% to 1.0% copper. Sites include Chuquicamata in Chile, Bingham Canyon Mine in Utah, United States and El Chino Mine in New Mexico, United States. According to the British Geological Survey in 2005, Chile was the top producer of copper with at least one-third world share followed by the United States, Indonesia and Peru. Copper can also be recovered through the in-situ leach process. Several sites in the state of Arizona are considered prime candidates for this method. The amount of copper in use is increasing and the quantity available is barely sufficient to allow all countries to reach developed world levels of usage.

The major applications of copper are electrical wire (60%), roofing and plumbing (20%), and industrial machinery (15%). Copper is used mostly as a pure metal, but when greater hardness is required, it is put into such alloys as brass and bronze (5% of total use). For more than two centuries, copper paint has been used on boat hulls to control the growth of plants and shellfish. A small part of the copper supply is used for nutritional supplements and fungicides in agriculture. Machining of copper is possible, although alloys are preferred for good machinability in creating intricate parts.

Words and terms

sterling silver　斯特林银

cupronickel　铜镍合金(白铜)

constantan　康铜

thermocouple　热电偶

azurite　蓝铜矿

malachite　孔雀石

turquoise　绿松石

pigment　颜料

verdigris　铜绿

bacteriostatic　抑菌剂

fungicide　杀真菌剂

preservative　防腐剂

respiratory enzyme　呼吸酶

cytochrome C oxidase　细胞色素 C 氧化酶

mollusc　软体动物

crustacean　甲壳动物

hemocyanin　血蓝蛋白

hemoglobin　血红蛋白

vertebrate　脊椎动物

macroscopic　肉眼可见的,宏观的

polycrystalline　多晶的

monocrystalline　单晶体

resistivity　电阻率

galvanic　原电池的

in-situ leach　原地浸出

porphyry　斑岩

Questions

Why is copper low in strength but good in electrical conductivity?

Why does the hull coating contain copper?

1.4.3　Magnesium

Magnesium is the ninth most abundant element in the universe. It is produced in large, aging stars from the sequential addition of three helium nuclei to a carbon nucleus. When such stars explode as supernovas, much of the magnesium is expelled into the interstellar medium where it may recycle into new star systems. Magnesium is the eighth most abundant element in the Earth's crust and the fourth most common element in the Earth (after iron, oxygen and silicon), making up 13% of the planet's mass and a large fraction of the planet's mantle. It is the third most abundant element dissolved in seawater, after sodium and chlorine.

Magnesium occurs naturally only in combination with other elements, where it invariably has a +2 oxidation state. The free element (metal) can be produced artificially, and is highly reactive (though in the atmosphere, it is soon coated in a thin layer of oxide that partly inhibits reactivity — see passivation). The free metal burns with a characteristic brilliant-white light. The metal is now obtained mainly by electrolysis of magnesium salts obtained from brine, and is used primarily as a component in aluminium-magnesium alloys, sometimes called magnalium or magnelium. Magnesium is less dense than aluminium, and the alloy is prized for its combination of lightness and strength.

Magnesium is the eighth-most-abundant element in the Earth's crust by mass and tied in seventh place with iron in molarity. It is found in large deposits of magnesite, dolomite, and other minerals, and in mineral waters, where magnesium ion is soluble. Although magnesium is found in more than 60 minerals, only dolomite, magnesite, brucite, carnallite, talc, and olivine are of commercial importance.

As of 2013, magnesium alloy consumption was less than one million tons per year, compared with 50 million tons of aluminum alloys. Its use has been historically limited by its tendency to corrode, creep at high temperatures, and combust.

The presence of iron, nickel, copper, and cobalt strongly activates corrosion. Greater than a very small percentage, these metals precipitate as intermetallic compounds, and the precipitate locales function as active cathodic sites that reduce water, causing the loss of magnesium. Controlling the quantity of these metals improves corrosion resistance. Sufficient manganese overcomes the corrosive effects of iron. This requires precise control over composition, increasing costs. Adding a cathodic poison captures atomic hydrogen within the structure of a metal. This prevents the formation of free hydrogen gas, an essential factor of corrosive chemical processes. The addition of about one in three hundred parts arsenic reduces its corrosion rate in a salt solution by a factor of nearly ten. Research showed that magnesium's tendency to creep at high-temperatures is eliminated by the addition of scandium and gadolinium. Flammability is greatly reduced by a small amount of calcium in the alloy.

(1) Applications

Magnesium is the third-most-commonly-used structural metal, following iron and aluminium. It is called the lightest useful metal by The Periodic Table of Videos. The main applications of magnesium are, in order: aluminium alloys, die-casting (alloyed with zinc), removing sulfur in the production of iron and steel, and the production of titanium in the Kroll process.

Magnesium is used in super-strong, lightweight materials and alloys. For example, when infused with silicon carbide nanoparticles, it has extremely high specific strength. Historically, magnesium was one of the main aerospace construction metals and was used for German military aircraft as early as World War Ⅰ (WW Ⅰ) and extensively for German aircraft in World War Ⅱ (WW Ⅱ).

The Germans coined the name "Elektron" for magnesium alloy, a term is still used today. In the commercial aerospace industry, magnesium was generally restricted to engine-related components, due fire and corrosion hazards. Currently, magnesium alloy used in aerospace is increasing, driven by the importance of fuel economy. Development and testing of new magnesium alloys continues, notably Elektron 21, which (in test) has proved suitable for aerospace engine, internal, and airframe components. The European Community runs three R&D magnesium projects in the Aerospace priority of Six Framework Program. In the form of thin ribbons, magnesium is used to purify solvents, for example, preparing super-dry ethanol.

Wright Aeronautical used a magnesium crankcase in the WW Ⅱ-era Wright Duplex Cyclone aviation engine. This presented a serious problem for the earliest models of the Boeing B-29 heavy bomber when an in-flight engine fire ignited the engine crankcase. The resulting combustion was as hot as 3,100 ℃ and could sever the wing spar from the fuselage.

(2) Mg alloy motorcycle engine blocks

Mercedes-Benz used the alloy Elektron in the body of an early model Mercedes-Benz 300 SLR; these cars ran (with successes) at Le Mans, the Mille Miglia, and other world-class race events in 1955. Porsche used magnesium alloy frames in the 917/053 that won Le Mans in 1971, and continues to use magnesium alloys for its engine blocks due to the weight advantage. Volkswagen Group has used magnesium in its engine components for many years.

BMW used magnesium alloy blocks in their N52 engine, including an aluminium alloy insert for the cylinder walls and cooling jackets surrounded by a high-temperature magnesium alloy AJ62A. The engine was used worldwide between 2005 and 2011 in various 1, 3, 5, 6, and 7 series models; as well as the Z4, X1, X3, and X5. Chevrolet used the magnesium alloy AE 44 in the 2006 Corvette Z06. Both AJ62A and AE44 are recent developments in high-temperature low-creep magnesium alloys. The general strategy for such alloys is to form intermetallic precipitates at the grain boundaries, for example by adding mischmetal or calcium. New alloy development and lower costs that make magnesium competitive with aluminium will increase the number of automotive applications.

Because of low weight and good mechanical and electrical properties, magnesium is widely used for manufacturing of mobile phones, laptop and tablet computers, cameras, and other electronic components.

Products made of magnesium: firestarter and shavings, sharpener, magnesium ribbon. Magnesium, being readily available and relatively nontoxic, has a variety of uses.

Magnesium is flammable, burning at a temperature of approximately 3,100 ℃, and the autoignition temperature of magnesium ribbon is approximately 473 ℃. It produces intense, bright, white light when it burns. Magnesium's high combustion temperature makes it a useful tool for starting emergency fires. Other uses include flash photography, flares, pyrotechnics, and fireworks sparklers. Magnesium is also often used to ignite thermite or other materials that require a high ignition temperature.

Magnesium firestarter (in left hand), used with a pocket knife and flint to create sparks, can ignite the shavings. In the form of turnings or ribbons, to prepare Grignard reagents, it is useful in organic synthesis. It can be an additive agent in conventional propellants and the production of nodular graphite in cast iron, a reducing agent to separate uranium and other metals from their salts, and a sacrificial (galvanic) anode to protect boats, underground tanks, pipelines, buried structures, and water heaters. Alloyed with zinc to produce the zinc sheet can be used in photoengraving plates in the printing industry, dry-cell battery walls, and roofing.

As a metal, this element's principal use is as an alloying additive to aluminium with these aluminium-magnesium alloys being used mainly for beverage cans, sports equipment such as golf clubs, fishing reels, and archery bows and arrows.

Special, high-grade car wheels of magnesium alloy are called "mag wheels", although the term is often misapplied to aluminium wheels. Many car and aircraft manufacturers have made engine and body parts from magnesium.

Magnesium batteries have been commercialized as primary batteries, and are an active topic of research for rechargeable secondary batteries.

Words and terms

supernova　超新星

interstellar　星际的

sodium　钠

brine　盐水

molarity　摩尔浓度

dolomite　白云石

magnesite　菱镁矿

brucite　红锌矿

carnallite　光卤石

talc　滑石

olivine　橄榄石

combust　燃烧

cathodic　阴极的

arsenic　砷

scandium　钪

gadolinium　钆

solvents　有机溶剂

ethanol　乙醇

crankcase　曲柄轴箱

bomber　轰炸机

fuselage　机身

mischmetal　混合稀土金属

flare　照明弹

pyrotechnics　烟花

thermite　铝热剂

Grignard reagents　格林试剂

anode　阳极

Questions

What limits the use of magnesium?

What do you think is the biggest feature of magnesium?

2

Heat treatment

Heat treating is a group of industrial and metalworking processes used to alter the physical, and sometimes chemical, properties of a material. The most common application is metallurgical. Heat treatments are also used in the manufacture of many other materials, such as glass. Heat treatment involves the use of heating or chilling, normally to extreme temperatures, to achieve a desired result such as hardening or softening of a material. Heat treatment techniques include annealing, case hardening, precipitation strengthening, tempering, normalizing and quenching. It is noteworthy that while the term heat treatment applies only to processes where the heating and cooling are done for the specific purpose of altering properties intentionally, heating and cooling often occur incidentally during other manufacturing processes such as hot forming or welding.

Metallic materials consist of a microstructure of small crystals called "grains" or crystallites. The nature of the grains (i. e. grain size and composition) is one of the most effective factors that can determine the overall mechanical behavior of the metal. Heat treatment provides an efficient way to manipulate the properties of the metal by controlling the rate of diffusion and the rate of cooling within the microstructure. Heat treating is often used to alter the mechanical properties of a metallic alloy, manipulating properties such as the hardness, strength, toughness, ductility, and elasticity.

There are two mechanisms that may change an alloy's properties during heat treatment: the formation of martensite causes the crystals to deform intrinsically, and the diffusion mechanism causes changes in the homogeneity of the alloy.

The crystal structure consists of atoms that are grouped in a very specific arrangement, called a lattice. In most elements, this order will rearrange itself, depending on conditions like temperature and pressure. This rearrangement, called allotropy or polymorphism, may occur several times, at many different temperatures for a particular metal. In alloys, this rearrangement may cause an element that will not normally dissolve into the base metal to suddenly become soluble, while a reversal of the allotropy will make the elements either partially or completely insoluble.

When in the soluble state, the process of diffusion causes the atoms of the dissolved element to spread out, attempting to form a homogenous distribution within the crystals of the base metal. If the alloy is cooled to an insoluble state, the atoms of the dissolved constituents (solutes) may migrate out of the solution. This type of diffusion, called precipitation, leads to nucleation, where the migrating atoms group together at the grain-boundaries. This forms a microstructure generally consisting of two or more distinct phases. Steel that has been cooled slowly, for instance, forms a laminated structure composed of alternating layers of ferrite and cementite, becoming soft pearlite. After heating the steel to the austenite phase and then quenching it in water, the microstructure will be in the martensitic phase. This is due to the fact that the steel will change from the austenite phase to the martensite phase after quenching. It should be noted that some pearlite or ferrite may be present if the quench did not rapidly cool off all the steel.

Unlike iron-based alloys, most heat treatable alloys do not experience a ferrite transformation. In these alloys, the nucleation at the grain-boundaries often reinforces the structure of the crystal matrix. These metals harden by precipitation. Typically a slow process, depending on temperature, this is often referred to as "age hardening".

Many metals and nonmetals exhibit a martensite transformation when cooled quickly (with external media like oil, polymer, water, etc.). When a metal is cooled very quickly, the insoluble atoms may not be able to migrate out of the solution in time. This is called a "diffusionless transformation." When the crystal matrix changes to its low temperature arrangement, the atoms of the solute become trapped within the lattice. The trapped atoms prevent the crystal matrix from completely changing into its low temperature allotrope, creating shearing stresses within the lattice. When some alloys are cooled quickly, such as steel, the martensite transformation hardens the metal, while in others, like aluminum, the alloy becomes softer.

Words and terms

anneal 退火

case harden 表面淬火

precipitation strengthen 析出强化

temper 回火

normalize 正火

quench 淬火

hot forming 热加工成形

allotropy 同素异形体

polymorphism 多晶型

nucleation 形核

Questions

What kind of conditions should be satisfied for heat treatment of materials?

What's the difference between heat treatment and hot forming?

2. 1   Effects of time and temperature

Time-temperature transformation (TTT) diagram for steel is shown in Fig. 2. 1. The black curves represent different cooling rates (velocity) when cooled from the upper critical (A3) temperature. V1 produces martensite. V2 has pearlite mixed with martensite. V3 produces bainite, along with pearlite and martensite.

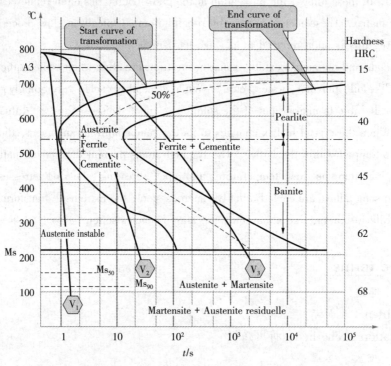

Fig. 2. 1   Time-temperature transformation (TTT) diagram for steel.

Proper heat treating requires precise control over temperature, time held at a certain temperature and cooling rate.

With the exception of stress-relieving, tempering, and aging, most heat treatments begin by heating an alloy beyond the upper transformation (A3) temperature. This temperature is referred to as an "arrest" because at the A3 temperature the metal experiences a period of hysteresis. At this point, all of the heat energy is used to cause the crystal change, so the temperature stops rising for a short time (arrests) and then continues climbing once the change is complete. Therefore, the alloy must be heated above the critical temperature for a transformation to occur. The alloy will usually be held at this temperature long enough for the heat to completely penetrate the alloy, thereby bringing it into a complete solid solution.

Because a smaller grain size usually enhances mechanical properties, such as toughness, shear strength and tensile strength, these metals are often heated to a temperature that is just above the upper critical-temperature, in order to prevent the grains of solution from growing too large. For instance, when steel is heated above the upper critical-temperature, small grains of austenite form. These grow larger as temperature is increased. When cooled very quickly, during a martensite transformation, the austenite grain-size directly affects the martensitic grain-size. Larger grains have large grain-boundaries, which serve as weak spots in the structure. The grain size is usually controlled to reduce the probability of breakage.

The diffusion transformation is very time-dependent. Cooling a metal will usually suppress the precipitation to a much lower temperature. Austenite, for example, usually only exists above the upper critical temperature. However, if the austenite is cooled quickly enough, the transformation may be suppressed for hundreds of degrees below the lower critical temperature. Such austenite is highly unstable and, if given enough time, will precipitate into various microstructures of ferrite and cementite. The cooling rate can be used to control the rate of grain growth or can even be used to produce partially martensitic microstructures. However, the martensite transformation is time-independent. If the alloy is cooled to the martensite transformation (Ms) temperature before other microstructures can fully form, the transformation will usually occur at just under the speed of sound.

When austenite is cooled slow enough that a martensite transformation does not occur, the austenite grain size will have an effect on the rate of nucleation, but it is generally temperature and the rate of cooling that controls the grain size and microstructure. When austenite is cooled extremely slow, it will form large ferrite crystals filled with spherical inclusions of cementite. This microstructure is referred to as "sphereoidite." If cooled a little faster, then coarse pearlite will form. Even faster, and fine pearlite will form. If cooled even faster, bainite will form. Similarly, these microstructures will also form if cooled to a specific temperature and then held there for a certain time.

Most non-ferrous alloys are also heated in order to form a solution. Most often, these are then cooled very quickly to produce a martensite transformation, putting the solution into a supersaturated state. The alloy, being in a much softer state, may then be cold worked. This cold working increases the strength and hardness of the alloy, and the defects caused by plastic deformation tend to speed up precipitation, increasing the hardness beyond what is normal for the alloy. Even if not cold worked, the solutes in these alloys will usually precipitate, although the process may take much longer. Sometimes these metals are then heated to a temperature that is below the lower critical (A1) temperature, preventing recrystallization, in order to speed-up the precipitation.

Words and terms

bainite　贝氏体

hysteresis　滞后现象

sphereoidite　球状渗碳体

recrystallization　再结晶

Questions

What is time-temperature transformation (TTT) diagram?

Can austenite exist at room temperature?

From the TTT diagram, what kind of characteristics can bainite have?

2.2　Typical heat treatments of steel

The purpose of heat treating carbon steel is to change the mechanical properties of steel, usually ductility, hardness, yield strength, or impact resistance. Note that the electrical and thermal conductivity are only slightly altered. As with most strengthening techniques for steel, Young's modulus (elasticity) is unaffected. All treatments of steel trade ductility for increased strength and vice versa. Iron has a higher solubility for carbon in the austenite phase; therefore, all heat treatments, except spheroidizing and process annealing, start by heating the steel to a temperature at which the austenitic phase can exist. The steel is then quenched (heat drawn out) at a moderate to low rate allowing carbon to diffuse out of the austenite forming iron-carbide (cementite) to precipitate leaving ferrite, or at a high rate, trapping the carbon within the iron thus forming martensite. The rate at which the steel is cooled through the eutectoid temperature (about 727 ℃) affects the rate at which carbon diffuses out of austenite and forms cementite. Generally speaking, cooling swiftly will leave iron carbide finely dispersed and produce a fine grained pearlite and cooling slowly will give a coarser pearlite. Cooling a hypoeutectoid steel (less than 0.77 $wt\%$ C) results in a lamellar-pearlitic structure of iron carbide layers with α-ferrite (nearly pure iron) between. If it is hypereutectoid steel (more than 0.77 $wt\%$ C) then the structure is full pearlite with small grains (larger than the pearlite lamella) of cementite formed on the grain boundaries. A eutectoid steel (0.77$wt\%$ C) will have a pearlite structure throughout the grains with no cementite at the boundaries. The relative amounts of constituents are found using the lever rule. The following (as shown in Fig 2.2) is a list of the types of heat treatments possible.

Spheroidizing: Spheroidite forms when carbon steel is heated to approximately 700 ℃ for over 30 hours. Spheroidite can form at lower temperatures but the time needed drastically increases, as this is a diffusion-controlled process. The result is a structure of rods or spheres of cementite within primary structure (ferrite or pearlite, depending on which side of the eutectoid you are on). The purpose is to soften higher carbon steels and allow more formability. This is the softest and most ductile form of steel. The image to the right shows where spheroidizing usually occurs.

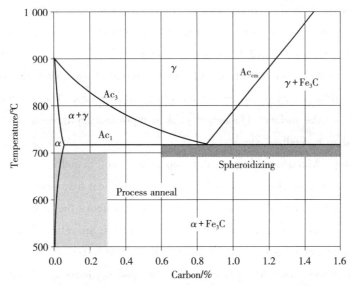

Fig. 2. 2　Iron-carbon phase diagram, showing the temperature and carbon ranges for certain types of heat treatments

Full annealing: Carbon steel is heated to approximately 40 ℃ above Ac_3 or Ac_m for 1 hour; this ensures all the ferrite transforms into austenite (although cementite might still exist if the carbon content is greater than the eutectoid). The steel must then be cooled slowly, in the realm of 20 ℃ per hour. Usually it is just furnace cooled, where the furnace is turned off with the steel still inside. This results in a coarse pearlitic structure, which means the "bands" of pearlite are thick. Fully annealed steel is soft and ductile, with no internal stresses, which is often necessary for cost-effective forming. Only spheroidized steel is softer and more ductile.

Process annealing: A process used to relieve stress in a cold-worked carbon steel with less than 0.3% C. The steel is usually heated to 550 ℃ to 650 ℃ for 1 hour, but sometimes temperatures are as high as 700 ℃. The image rightward shows the area where process annealing occurs.

Isothermal annealing: It is a process in which hypoeutectoid steel is heated above the upper critical temperature. This temperature is maintained for a time and then reduced to below the lower critical temperature and is again maintained. It is then cooled to room temperature. This method eliminates any temperature gradient.

Normalizing: Carbon steel is heated to approximately 55 ℃ above Ac_3 or Ac_m for 1 hour; this ensures the steel completely transforms to austenite. The steel is then air-cooled, which is a cooling rate of approximately 38 ℃ per minute. This results in a fine pearlitic structure, and a more-uniform structure. Normalized steel has a higher strength than annealed steel; it has a relatively high strength and hardness.

Quenching：Carbon steel with at least 0. 4 *wt%* C is heated to normalizing temperatures and then rapidly cooled (quenched) in water, brine, or oil to the critical temperature. The critical temperature is dependent on the carbon content, but as a general rule is lower as the carbon content increases. This results in a martensitic structure; a form of steel that possesses a super-saturated carbon content in a deformed body-centered cubic (BCC) crystalline structure, properly termed body-centered tetragonal (BCT), with much internal stress. Thus quenched steel is extremely hard but brittle, usually too brittle for practical purposes. These internal stresses mat cause stress cracks on the surface. Quenched steel is approximately three to four (with more carbon) fold harder than normalized steel.

Martempering (**Marquenching**)：Martempering is not actually a tempering procedure, hence the term is "marquenching". It is a form of isothermal heat treatment applied after an initial quench, typically in a molten salt bath, at a temperature just above the "martensite start temperature". At this temperature, residual stresses within the material are relieved and some bainite may be formed from the retained austenite which did not have time to transform into anything else. In industry, this is a process used to control the ductility and hardness of a material. With longer marquenching, the ductility increases with a minimal loss in strength; the steel is held in this solution until the inner and outer temperatures of the part equalize. Then the steel is cooled at a moderate speed to keep the temperature gradient minimal. Not only does this process reduce internal stresses and stress cracks, but it also increases the impact resistance.

Tempering：This is the most common heat treatment encountered, because the final properties can be precisely determined by the temperature and time of the tempering. Tempering involves reheating quenched steel to a temperature below the eutectoid temperature then cooling. The elevated temperature allows very small amounts of spheroidite to form, which restores ductility, but reduces hardness. Actual temperatures and times are carefully chosen for each composition.

Austempering：The austempering process is the same as martempering, except the quench is interrupted and the steel is held in the molten salt bath at temperatures between 205 ℃ and 540 ℃, and then cooled at a moderate rate. The resulting steel, called bainite, produces an acicular microstructure in the steel that has great strength (but less than martensite), greater ductility, higher impact resistance, and less distortion than martinsite steel. The disadvantage of austempering is that it can only be used on a few steels, and it requires a special salt bath.

Words and terms

vice versa *反之亦然*

spheroidizing *球状化退火*

iron-carbide *碳化铁(渗碳体)*

isothermal annealing *等温退火*

temperature gradient 温度梯度
martempering 马氏体等温淬火
austempering 奥氏体等温淬火
acicular 针状

Questions

What is the function of annealing?

What's the difference between normalizing and annealing?

What are the sources of residual stresses in hardened steel?

3

Metal casting

❦❦❦

3.1 Introduction

A casting may be defined as a "metal object obtained by allowing molten metal to solidify in a mold", the shape of the object being determined by the shape of the mold cavity. Casting is basically melting a solid material, heating to a special temperature, and pouring the molten material into a cavity or mold, which is in proper shape. Casting has been known by human being since the 4th century B. C. In metalworking, casting means a process, in which liquid metal is poured into a mold, which contains a hollow cavity of the desired shape, and then allowed to cool and solidify. The solidified part is also known as a casting, which is ejected or broken out of the mold to complete the process. Casting is most often used for making complex shapes that would be difficult or uneconomical to make by other methods. Casting processes have been known for thousands of years, and widely used for sculpture, especially in bronze, jewelry in precious metals, and weapons and tools. Traditional techniques include lost-wax casting, plaster mold casting and sand casting. The modern casting process is subdivided into two main categories: expendable and non-expendable casting. It is further broken down by the mold material, such as sand or metal, and pouring method, such as gravity, vacuum, or low pressure.

Words and terms

casting 铸造
solid material 固体材料
metalworking 金属加工
molten metal 熔融金属
temperature 温度
mold 模具

Questions

Identify the metal casting.

What are two advantages and one disadvantage of the permanent molding?

What are two main categories of the modern casting process?

3.2 Casting procedure

In all casting processes six basic factors In all casting processes are involved. These are as follows:

1. A mold cavity, having the desired shape and size and with due allowance for shrinkage of the solidifying metal, must be produced. Any complexity of shape desired in the finished casting must exist in the cavity. Consequently, the mold material must be such as to reproduce the desired detail and also have a refractory character so that it will not be significantly affected by the molten metal that it contains. Either a new mold must be prepared for each casting, or it must be made from a material that can withstand being used for repeated castings, the latter being called permanent molds.

2. A suitable means must be available for melting the metal that is to be cast, providing not only adequate temperature, but also satisfactory quality and quantity at low cost.

3. The molten metal must be introduced into the mold in such a manner that all air or gases in the mold, prior to pouring or generated by the action of the hot metal upon the mold, will escape, and the mold will be completely filled. A quality casting must be dense and free from defects such as air holes.

4. Provision must be made so that the mold will not cause too much restraint to the shrinkage that accompanies cooling after the metal has solidified. Otherwise, the casting will crack while its strength is low. In addition, the design of the casting must be such that solidification and solidification shrinkage can occur without producing cracks and internal porosity or voids.

5. It must be possible to remove the casting from the mold so a permanent mold must be made in two or more sections.

6. After removal from the mold, finishing operations may need to be performed to remove extraneous material that is attached to the casting as the result of the method of introducing the metal into the cavity, or is picked up from the mold through contact with the metal.

Casting processes are important and extensively used manufacturing methods, enabling the production of very complex or intricate parts in nearly all types of metals with high production rates, average to good tolerances and surface roughness, and good material properties. The competitiveness of the casting processes is based primarily on the fact that casting allows the elimination of substantial amounts of expensive machining often required in alterative production methods.

As mentioned, many different casting processes have been developed. The names associated with the processes may be related to the type of mold (nonpermanent, permanent) or to the mold material or the pouring method (gravity, high pressure, low pressure). Furthermore, the application of the names is not always consistent, which sometimes causes confusion. Table 3.1 shows the major casting processes classified according to the different characteristics. The most commonly used names are given, but if doubt about them arises, they can be identified by their characteristics. The individual processes are described later.

Tabel 3.1 **Some characteristics of major casting processes**

Type of mold	Mold material	Pouring principle	Pattern material	Process name	Grouping
Nonpermantent (single-purpose)	Sand (green)	Gravity	Wood, metal. plastics	Green sand, dry sand core sand casting	Sand casting
Permanent	Alloy steels Graphite, Steel Cast iron	High pressure	—	Die casting	Permanent (metallic) mold casting
		Low pressure	—	Low pressure (permanent mold) casting	
		Gravity	—	Non-Pressure-gravity permanent mold casting	
Nonpermantent (single-purpose)	Nonmetallic (sand, plaster, ceramics,etc.)	Gravity (Low pressure)	Metal, wax, plastic, rubber	Shell mold cast	Precision casting
				Plaster mold casting	
				Ceramic shell mold casting	
			Wax	"Lost wax" casting (investment casting)	
Nonpermantent permanent	Nonmetallic Metallic	Centrifugal forces	—	Centrifugal casting	Centrifugal casting

Words and terms

mold cavity　模具型腔

solidification　凝固

surface roughness　表面粗糙度

air holes　气孔

manufacture　加工;制造

high pressure　高压

Questions

Identify and list the various types of casting processes.

Name two advantages and one disadvantage of permanent molding.

What are the characteristics of major casting processes?

4

Die for metal forming

4.1　Single die

A wide variety of dies are used for sheet metal forming (including cutting). In single operation processes, there are blanking, punching, bending, drawing flanging dies, and the like. In multiple-operation processes, there are compound, combination, progressive, transfer dies, and the like.

4.1.1　Blanking dies

The various types of die cutting include blanking, punching, perforating, parting, shaving, trimming, etc. Among them the blanking is the basic die cutting. The two principal types of blanking dies are designated as "drop-through" and "return-blank" designs. In a drop-through die, the blank drops through the die block and onto the bolster plate, often provided with an unloading chute, or drop through a hole in both die block and bolster plate. In a return-blank die, the blank is pushed up into the die cavity, after that it is ejected by a spring-loaded knockout. Fig. 4. 1 shows a drop-through blanking die.

Words and terms

　　metal forming　金属成形
　　bolster　支撑
　　plate　平面，板材
　　blanking dies　冲模
　　material　材料
　　design　设计

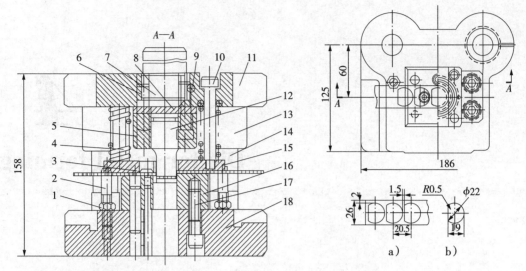

Notes: 1-nut; 2-stock guide bar; 3-fixed stop; 4-spring; 5-punch retainer; 6,9-pins; 7-dieshank; 8-punch back up plate;
10,17-screw; 11-upper shoe; 12-punch; 13-guide -pin bushing; 14-guide -pin; 15-solid stripper;
16-die;18-lower shoe

Fig. 4. 1　Drop - through blanking die. a) layout (material st 08, thickness 2 mm). b) workpiece

Questions

Identify the single die.

Declare the various types of die cutting.

4.1.2　Bending dies

Bending dies are used on presses for bending sheet metal or wire parts into various shapes. The work is done by the punch pushing the stock into cavities or depressions of similar shape in the die or by auxiliary attachments operated by the descending punch. The basic types of bending applicable to sheet metal forming are straight bending, flanging bending, and contour bending. The bending die shown in Fig. 4. 2, used in a conventional press, was designed to form a V-shaped part.

4.1.3　Drawing dies

Drawing is a process of cold forming a flat precut metal blank into a hollow vessel without excessive wrinkling, thinning or fracturing. The various forms produced may be cylindrical or box doped, with straight or tapered sides or a

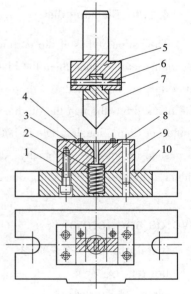

Notes: 1-screw; 2-spring; 3-ejector; 4-fixed
stop; 5-die shank; 6,9-pins; 7-punch;
8-die; 10-lower shoe

Fig. 4. 2　V -bend die.

combination of strait, tapered, and curved sides. When the height of the cup drawn from the blank is larger than the diameter of the required part, the cup must be redrawn, according to recommended percentages, until the required diameter is obtained.

The drawing dies can be classified, in a simplified manner, into the following types: single-action die with spring, rubber, or air-cushion blankholding, which is operated by a single-action press, double-action die operated by a double-action press with the blankholder attached to the outer press slide; direct redrawing die in a double-action press; inverted redrawing die in a double-action press; etc. Fig. 4.3 shows a drawing die, with rubber blankholding, used in a single-action press. To prevent the flange from buckling, a blankholder is used and the clamping force will be of the same order as the punch force. In this die, the blankholder 1 not only plays a part in blankholding, but also in ejecting and positioning.

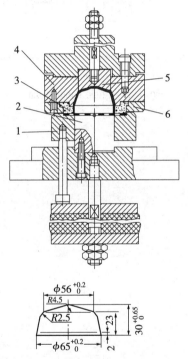

Notes: 1-blankholder; 2-punch; 3-die section(1); 4-die section (2); 5-knock-out plate; 6-die retainer

Fig. 4.3　Drawing die with rubber blankholder

4.1.4　Hole flanging dies

Besides bending and drawing dies, there are many forming dies: such as flanging, embossing, curling, hemming, etc. Flanging is similar to the bending of sheet metal, except that during flanging, the bent down metal is short compared to overall part size. Fig. 4.4 illustrates a hole flanging die.

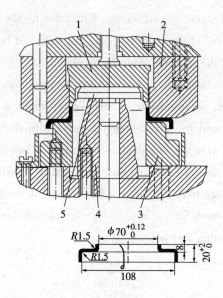

Notes: 1-knockout plate; 2-die; 3-pressure plate;

4-punch; 5-positioning plate

Fig. 4. 4   A hole flanging die

Words and terms

bend   弯曲

wrinkling   起皱

blankholder   防皱压板

embossing   压边

drawing   拉拔

thinning   减薄

flanging   卷边

Questions

What can bending dies be used for?

Depict the process of drawing.

Which can prevent the flange from buckling?

4.2 Compound dies

The compound die is single-station die, the elements of which are designed around a common centerline (usually vertical) and in which two or more operations are completed during a single press stroke, but only cutting operations are done. The compound blank punch die used for nesting cutting three spacers simultaneously is shown in Fig. 4.5.

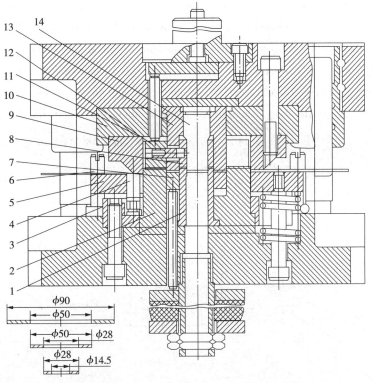

Notes: 1-lower, inner punch-die; 2-lower, outer punch-die; 3-lower die retainer; 4-movable stop; 5-movable stripper; 6-stock huide bar; 7-kicker; 8-inner ejector; 9-die; 10-upper die retainer; 11-connecting pin; 12-outer ejector; 13-upper punch die; 14-punch

Fig. 4.5　Compound blank - punch die

4.3　Progressive dies

Progressive operation are multiple-station operations performed by means of a die having several stations, each of which performs a different operation as the stock passes through the die. The design may include components or devices to position, locate, or guide the stock. Idle station, at which no work is performed, is used to spread out closely spaced operations or to better distribute the forces required performing the work. In progressive operations, parts are connected by carrier strip until the final parting or cutoff operation. Individual operations performed in a progressive die are usually relatively simple; when they are combined in several stations, however, the most practical and economical strip design for optimum operations of the die often becomes difficult to devise.

A progressive die used for punching and blanking is shown in Fig. 4.6. It can be seen that there is an idle station which is used to strengthen die blocks, stripper plate, punch and die retainers; there is a pitch of 52 mm long, which must be kept constant between successive stations. Progressive dies are similar in function to compound dies in that they combine in one die set several operations that are performed with one stroke of the press. In a progressive die, however, the operations are separated and distributed among a number of stations. The stock progresses through these stations in the strip form until the finished workpiece is cut from the strip at the last station. Consequently, at the start of a strip, there are several press strokes before a piece is produced. Thereafter, a finished workpiece is produced with each stroke of the press, up to the end of the strip. If the strips are short, sheared from sheets, they can be fed into the die so that each

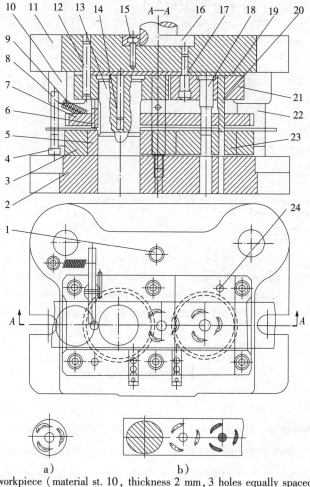

a) workpiece (material st. 10, thickness 2 mm, 3 holes equally spaced).

b) progressive strip layout.

Notes: 1-limited post; 2-lower shoe; 3-die retainer; 4-nut; 5-blanking die (block); 6-movable stop;
7-fixed stripper; 8, 15, 17-screw; 9-pull spring; 10-guide pin bushing; 11-upper shoe;
12, 24-pins; 13-blanking punch; 14-pilot; 16-die shank; 18, 20-piecing punches; 19-punch back up plate; 21-punch retainer; 22-guide pin; 23-piercing die (block)

Fig. 4. 6　progressive punch - blank die

strip is butted against the end of the preceding strip for continuous production. Otherwise, starting stops are used for each new strip. Coil stock is fed into the die at one end. A hole or notch is usually made in the strip at the first station, and subsequent stations use this as a pilot to keep the strip properly aligned and positioned while the operations are performed. When the workpiece is complete, it is cut from the strip, which has acted as a holder to carry the piece from station to station. Intricate pieces, often needing no further work, can be made in one press. In planning the strip layout for progressive-die operations, consideration must be given to development of the part outline, to provision for piloting, to distribution of press load and strength of die elements, and to ensuring minimum metal waste.

Words and terms

compound die   复合模

device   设备

cutoff   中止

centerline   中心线

progressive operation   连续操作

coil stock   卷料

Questions

Define compound die.

State briefly how compound die works.

Explain how progressive punch - blank die works.

5

Sand casting

❧❧❧

Casting is used to make metal products of almost any desired shape by the pouring of molten metal into a reshaped hollow mold. As the metal freezes, the mold is removed. This technique was learned thousands of years ago when it was discovered that damp sand could be packed by hand into almost any shape.

Generally speaking, clean, fine sand is placed in a wooden or steel box and packed around a preformed wooden pattern or actual object. When the pattern is removed, its imprint remains in the sand. Molten metal is then poured into the hollow mold. Castings made from sand molds have a rough surface. They must be cleaned, trimmed and machined at times. Sand molds must be repacked after each casting to obtain additional parts. Therefore, the casting of many pieces using sand molds requires much time and labor.

The procedure for making a typical sand mould is described in following steps.

First a bottom board is placed either on the moulding platform or on the floor, making the surface event. The drag-moulding flask is kept upside down on the bottom board along with the drag part of the pattern at the centre of the flask on the board. There should be enough clearance between the pattern and the walls of the flask which should be of the order of 50 mm to 100 mm. Dry facing sand is sprinkled over the board and pattern to provide a non-sticky layer. Freshly prepared moulding sand of requisite quality is now poured into the drag and on the pattern to a thickness of 30 mm to 50 mm. The rest of the drag flask is completely filled with the backup sand and uniformly rammed to compact the sand. The ramming of sand should be done properly so as not to compact it too hard, which makes the escape of gases difficult, nor too loose so that mould would not have enough strength. After the ramming is over, the excess sand in the flask is completely scraped using a flat bar to the level of the flask edges.

Now, with a vent wire, which is a wire of 1 mm to 2 mm diameter with a pointed end, vent holes are made in the drag to the full depth of the flask as well as the pattern to facilitate the removal of gases during casting solidification. This completes the preparation of the drag.

The finished drag flask is now rolled over the bottom board exposing the pattern. Using a slick, the edge of sand around the pattern is repaired and cope half of the pattern is placed over the drag pattern, aligning it with the help of dowel pins. The cope flask on top of the drag is located aligning again with the help of the pins. The dry parting sand is sprinkled all over the drag and on the pattern.

A sprue pin for making the sprue passage is located at a small distance of 50 mm from the pattern. Also a riser pin if required is kept at an appropriate place and freshly prepared moulding sand similar to that of the drag along with the backing sand is sprinkled. The sand is thoroughly rammed, excess sand scraped and vent holes are made all over in the cope as in the drag.

The sprue pin and the riser pin are carefully withdrawn from the flask. Later the pouring basin is cut near the top of the sprue. The cope is separated from the drag and any loose sand on the cope and drag interface of the drag is blown off with the help of bellows. Now the cope and the drag pattern halves are withdrawn by using the draw spikes and rapping the pattern all around to slightly enlarge the mould cavity so that the mould walls are not spoiled by the withdrawing pattern. The runners and the gates are cut in the mould carefully without spoiling the mould. Any excess or loose sand found in the runners and mould cavity is blown away using the bellows. Now the facing sand in the form of a paste is applied all over the mould cavity and the runners, which would give the finished casting a good surface finish.

A dry sand core is prepared using a core box. After suitable baking, it is placed in the mould cavity as shown in Fig. 5.1. The cope is replaced on the drag taking care of the alignment of the two means of the pins. After moulding, melting is the major factor which controls the quality of the casting. There are a number of methods available for melting foundry alloys such as pit furnace, open-hearth furnace, rotary furnace, cupola furnace, etc. The choice of the furnace depends on the amount and the type of alloy being melted. For melting cast iron, cupola in its various forms is extensively used basically because of its lower initial cost and lower melting cost.

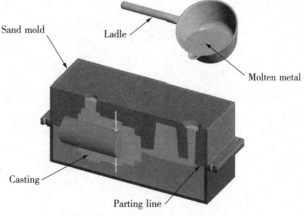

Fig. 5.1　Sand casting overview

Words and terms

hollow mold　空模

sand mould　砂模

metal freezes　金属凝固

casting solidification　铸件凝固

open-hearth furnace　平炉

cupola furnace　化铁炉；冲天炉

damp sand　湿砂

drag-moulding　拖模

poured into　倒入

pit furnace　井式炉；坑式炉

rotary furnace　旋转加热炉；回转炉

Questions

Explain the sand casting and its process.

Describe the characteristics of sand casting.

Describe the equipment involved in the sand casting process.

6

Permanent mould casting

Permanent mold casting is a metal casting process that employs reusable molds ("permanent molds"), usually made from metal. The most common process uses gravity to fill the mold, however gas pressure or a vacuum are also used. A variation on the typical gravity casting process, called slush casting, produces hollow castings. Common casting metals are aluminium, magnesium, and copper alloys. Other materials include tin, zinc, and lead alloys and iron and steel are also cast in graphite molds.

Fig. 6.1   Permanent mold casting

In all the processes that have been covered so far, a mould need to be prepared for each of the casting produced. For large scale production, making a mould for every casting to be produced maybe difficult and expensive. Therefore, a permanent mould, called "die" may be made from which a large number of castings, anywhere between 100 to 250,000 can be produced, depending on alloy used and the complexity of the casting. This process is called permanent mould casting or gravity die casting, since the metal enters the mould under gravity, as shown in Fig 6.1.

The mould material is selected on the consideration of the pouring temperature, size of the casting and frequency of the casting cycle. They determine the total heat to be borne by the die. Fine-grained grey cast iron is the most generally used die material. Alloy cast iron, C20 steel and alloy steel (H11 and H14) are also used for very large volumes and large parts. Graphite mould may be used for small volume production from aluminum and magnesium. The die life is less for higher melting temperature alloys such as copper or grey cast iron.

For making any hollow portions, cores are also used in permanent mould casting. The cores can be made out of metal, or sand. When sand cores are used, the process is called semi-permanent moulding. The metallic core cannot be complex with under-cuts and the like. Also, the metallic core is to be withdrawn immediately after solidification, otherwise, its extraction becomes difficult because of shrinkage. For complicated shapes, collapsible metal cores (multiple-piece cores) are sometimes used in permanent moulds. Their use is not extensive because of the fact that it is difficult to securely position the core as a single piece as also due to the dimensional variations that are likely to occur. Hence, with collapsible cores, the designer has to provide coarse tolerance on these dimensions.

Permanent mold casting is typically used for high-volume production of small, simple metal parts with uniform wall thickness. Non-ferrous metals are typically used in this process, such as aluminum alloys, magnesium alloys, and copper alloys. However, irons and steels can also be cast using graphite molds. Common permanent mold parts include gears and gear housings, pipe fittings, and other automotive and aircraft components such as pistons, impellers, and wheels.

The permanent mold casting process consists of the following steps:

1) Mold preparation. First, the mold is pre-heated to around 300 °F to 500 °F (150 °C to 260 °C) to allow better metal flow and reduce defects. Then, a ceramic coating is applied to the mold cavity surfaces to facilitate part removal and increase the mold lifetime.

2) Mold assembly. The mold consists of at least two parts, the two mold halves and any cores used to form complex features. Such cores are typically made from iron or steel, but expendable sand cores are sometimes used. In this step, the cores are inserted and the mold halves are clamped together.

3) Pouring. The molten metal is poured at a slow rate from a ladle into the mold through a sprue at the top of the mold. The metal flows through a runner system and enters the mold cavity.

4) Cooling. The molten metal is allowed to cool and solidify in the mold.

5) Mold opening. After the metal has solidified, the two mold halves are opened and the casting is removed.

6) Trimming. During cooling, the metal in the runner system and sprue solidify attached to the casting. This excess material is now cut away.

The mould cavity should normally, be simple without any undesirable drafts or undercuts, which interfere with the ejection of the solidified castings. In designing the permanent moulds, care should be taken to see that progressive solidification towards the riser is achieved. If the casting has heavy sections, which are likely to interfere with the progressive solidification, mould section around that area may be made heavier around that area to extract more heat. Chills supported by heavy air blast may be also used to remove the excess heat. Alternatively, cooling channels may be provided at the necessary points to get proper temperature distribution. The likely problems with the cooling water circulation are the formation of scales inside the cooling channels and their subsequent blocking after some use, as shown in Fig 6. 2.

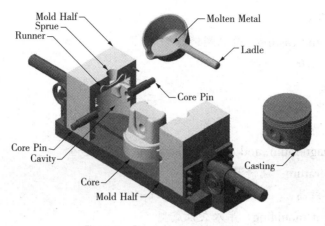

Fig. 6. 2　Permanent mold casting

The gating and risering system used are very similar to that of the sand casting. In fact, to get the proper gating arrangements, it may be desirable first to experiment with various gating systems in sand casting and then finally arrive at the correct gating system for the metallic mould.

The moulds are coated with a refractory material to a thickness of around 0. 8 mm. The coatings are used to increase the mould life by preventing the soldering of metal to the mould, by minimizing the thermal shock to the mould material, and by controlling the rate and direction of the casting solidification.

The coatings normally are mixtures of sodium silicate, kaolin clay, soapstone and talc. The coatings are both insulating type and lubricating type. The main requirement of a coaling is that it should be inert to the casting alloy. The coating may be applied by spraying or brushing. It must be thick enough to fill up any surface imperfections. Ting cycle, the temperature at which the mould is used depends on the pouring temperature. Coatings can be applied thicker at surfaces, which need to be cooled slowly, for example, sprues, runners, risers and thin sections. The maximum thickness of a coating required is about 0. 8 mm.

Under the regular casting cycle, the temperature at which the mould is used depends on the pouring temperature, casting cycle frequency, casting weight, casting shape, casting wall thickness, wall thickness of the mould and the thickness of the mould coating. If the casting is done with the cold die, the first few casting are likely to have miss-runs till the die reaches its operating temperature. To avoid this, the mould should be preheated to its operating temperature, preferably in an oven.

The materials, which are normally cast in permanent moulds, are aluminum alloys, magnesium alloys, copper alloys, zinc alloys, and grey cast iron. The sizes of castings are limited to 15 kg in most of the materials. But in case of aluminum, large castings with a mass of up to 350 kg have been produced. Permanent mould casting is particularly suited to high volume production of small, simple castings with uniform wall thickness and no intricate details.

Words and terms

permanent mould casting　金属型铸造

gas pressure　气压

vacuum　真空

variation　变化,变动

slush casting　空壳铸件

aluminium, magnesium, and copper alloys　铝,镁,铜合金

pouring temperature　浇铸温度

fine-grained　细晶

semi-permanent moulding　半永久性成型

metallic core　金属芯

gating and risering system　浇冒口系统

sodium silicate　硅酸钠;水玻璃

kaolin clay　高岭土

Questions

Describe the process of permanent mould casting.

How to fix pouring temperature?

Describe briefly the application of permanent mould casting.

7

Investment casting

In investment casting (also called "lost wax casting" or "precision casting") a pattern of wax is used. Investment casting has been used in various forms for the last 5,000 years. In its earliest forms, beeswax was used to form patterns necessary for the casting process. Today, more advanced waxes, refractory materials and specialist alloys are typically used for making patterns. Investment casting is valued for its ability to produce components with accuracy, repeatability, versatility and integrity in a variety of metals and high-performance alloys.

Fig. 7. 1 Products processed by investment casting

The fragile wax patterns must withstand forces encountered during the mold making. Much of the wax used in investment casting can be reclaimed and reused. Lost-foam casting is a modern form of investment casting that eliminates certain steps in the process.

Investment casting derives its name from the pattern being invested with a refractory material. Many materials are suitable for investment casting; examples are stainless steel alloys, brass, aluminum, carbon steel and glass, as shown in Fig 7. 1. The material is poured into a cavity in a refractory material that is an exact duplicate of the desired part. Due to the hardness of refractory materials used, investment casting can produce products with exceptional surface qualities, which can reduce the need for secondary machine processes.

Investment casting requires the use of a metal die, wax, ceramic slurry, furnace, molten metal, and any machines needed for sandblasting, cutting, or grinding. The process steps are as follows, as shown in Fig 7. 2.

1）Pattern creation. The wax patterns are typically injection molded into a metal die and are formed as one piece. Cores may be used to form any internal features on the pattern. Several of these patterns are attached to a central wax gating system (sprue, runners, and risers), to form a tree-like assembly. The gating system forms the channels through which the molten metal will flow to the mold cavity.

2）Mold creation. This "pattern tree" is dipped into a slurry of fine ceramic particles, coated with more coarse particles, and then dried to form a ceramic shell around the patterns and gating system. This process is repeated until the shell is thick enough to withstand the molten metal it will encounter. The shell is then placed into an oven and the wax is melted out leaving a hollow ceramic shell that acts as a one-piece mold, hence the name "lost wax" casting.

3）Pouring. The mold is preheated in a furnace to approximately 1,000 ℃ (1,832 ℉) and the molten metal is poured from a ladle into the gating system of the mold, filling the mold cavity. Pouring is typically achieved manually under the force of gravity, but other methods such as vacuum or pressure are sometimes used.

4）Cooling. After the mold has been filled, the molten metal is allowed to cool and solidify into the shape of the final casting. Cooling time depends on the thickness of the part, thickness of the mold, and the material used.

5）Casting removal. After the molten metal has cooled, the mold can be broken and the casting removed. The ceramic mold is typically broken using water jets, but several other methods exist. Once removed, the parts are separated from the gating system by either sawing or cold breaking (using liquid nitrogen).

6）Finishing. Often times, finishing operations such as grinding or sandblasting are used to smooth the part at the gates. Heat treatment is also sometimes used to harden the final part.

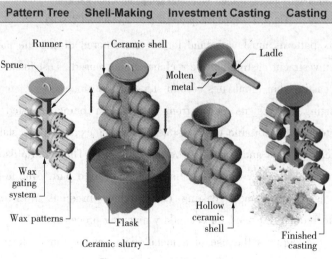

Fig. 7. 2　Investment casting

Investment casting derives its name from the pattern being invested (surrounded) with a refractory material. Many materials are suitable for investment casting; examples are stainless steel alloys, brass, aluminium, carbon steel and glass. The material is poured into a cavity in a refractory material that is an exact duplicate of the desired part. Due to the hardness of refractory materials used, investment casting can produce products with exceptional surface qualities, which can reduce the need for secondary machine processes.

The process can be used for both small castings of a few ounces and large castings weighing several hundred pounds. It can be more expensive than die casting or sand casting, but per-unit costs decrease with large volumes. Investment casting can produce complicated shapes that would be difficult or impossible with other casting methods. It can also produce products with exceptional surface qualities and low tolerances with minimal surface finishing or machining required.

Advantages of investment casting

◇ Excellent surface finish

◇ High dimensional accuracy

◇ Extremely intricate parts are castable

◇ Almost any metal can be cast

◇ No flash or parting lines

Disadvantages of investment casting

The main disadvantage is the overall cost, especially for short-run productions. Some of the reasons for the high cost include specialized equipment, costly refractories and binders, many operations to make a mould, a lot of labor is needed and occasional minute defects. However, the cost is still less than producing the same part by machining from bar stock; for example, gun manufacturing has moved to investment casting to lower costs of producing pistols.

◇ It can be difficult to cast objects requiring cores.

◇ This process is expensive, is usually limited to small casting, and presents some difficulties where cores are involved.

◇ Holes cannot be smaller than 1/16 inch (1.6 mm) and should be no deeper than about 1.5 times the diameter.

◇ Investment castings require longer production cycles compared to other casting process.

Words and terms

investment casting　熔模铸造

lost wax casting　蜡模铸造;失蜡铸造

precision casting　精密铸造

beeswax　蜂蜡

refractory materials　耐火材料

water jets　水注[流];喷水式推进器

heat treatment　热处理
liquid nitrogen　液氮;液化氮;液态氮
pistols　手枪
binders　刹车器

Questions

Describe briefly the characteristics and process of investment casting.
What problem does investment casting need to pay attention to?

8

Die casting

Die casting equipment was invented in 1838 for the purpose of producing movable type for the printing industry. The first die casting-related patent was granted in 1849 for a small hand-operated machine for the purpose of mechanized printing type production. In 1885 Otto Mergenthaler invented the linotype machine, an automated type-casting device which became the prominent type of equipment in the publishing industry. The Soss die-casting machine, manufactured in Brooklyn, NY, was the first machine to be sold in the open market in North America. Other applications grew rapidly, with die casting facilitating the growth of consumer goods and appliances by making affordable the production of intricate parts in high volumes. In 1966, General Motors released the accrued process.

Die casting is a manufacturing process that can produce geometrically complex metal parts through the use of reusable molds, called dies, as shown in Fig 8.1. The die casting process involves the use of a furnace, metal, die casting machine, and die. The metal, typically a non-ferrous alloy such as aluminum or zinc, is melted in the furnace and then injected into the dies in the die casting machine. There are two main types of die casting machines - hot chamber machines (used for alloys with low melting temperatures, such as zinc) and cold chamber machines (used for alloys with high melting temperatures, such as aluminum). The differences between these machines will be detailed in the sections on equipment and tooling. However, in both machines, after the molten metal is injected into the dies, it rapidly cools and solidifies into the final part, called the casting. The steps in this process are described in greater detail in the next section.

The castings that are created in this process can vary greatly in size and weight, ranging from a couple ounces to 100 pounds. One common application of die cast parts are housings - thin-walled enclosures, often requiring many ribs and bosses on the interior. Metal housings for a variety of appliances and equipment are often die cast. Several automobile components are also manufactured using die casting, including pistons, cylinder heads, and engine blocks. Other common die cast parts include propellers, gears, bushings, pumps, and valves.

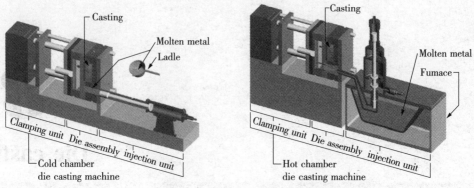

Fig. 8. 1 Die casting cold and hot chamber machine overview

Words and terms

die casting 拉模铸造
equipment 设备
printing industry 印刷工业
hand-operated machine 手工操作的机器
reusable molds 可重复使用的模具
non-ferrous alloy 有色金属合金;有色合金
interior 内部
propellers 螺旋桨,推进器
gears 挡;齿轮;装备

Questions

Describe briefly the history of die casting.

What is the application of die casting?

8.1 Process

1) Clamping. The first step is the preparation and clamping of the two halves of the die. Each die half is first cleaned from the previous injection and then lubricated to facilitate the ejection of the next part. The lubrication time increases with part size, as well as the number of cavities and side-cores. Also, lubrication may not be required after each cycle, but after 2 or 3 cycles, depending upon the material. After lubrication, the two die halves, which are attached inside the die casting machine, are closed and securely clamped together. Sufficient force must be applied to the die to keep it securely closed while the metal is injected. The time required to close and clamp the die is dependent upon the machine - larger machines (those with greater clamping forces) will

require more time. This time can be estimated from the dry cycle time of the machine.

2) Injection. The molten metal, which is maintained at a set temperature in the furnace, is next transferred into a chamber where it can be injected into the die. The method of transferring the molten metal is dependent upon the type of die casting machine, whether a hot chamber or cold chamber machine is being used. The difference in this equipment will be detailed in the next section. Once transferred, the molten metal is injected at high pressures into the die. Typical injection pressure ranges from 7 to 138 Mpa. This pressure holds the molten metal in the dies during solidification. The amount of metal that is injected into the die is referred to as the shot. The injection time is the time required for the molten metal to fill all of the channels and cavities in the die. This time is very short, typically less than 0. 1 s, in order to prevent early solidification of any one part of the metal. The proper injection time can be determined by the thermodynamic properties of the material, as well as the wall thickness of the casting. A greater wall thickness will require a longer injection time. In the case where a cold chamber die casting machine is being used, the injection time must also include the time to manually ladle the molten metal into the shot chamber.

3) Cooling. The molten metal that is injected into the die will begin to cool and solidify once it enters the die cavity. When the entire cavity is filled and the molten metal solidifies, the final shape of the casting is formed. The die cannot be opened until the cooling time has elapsed and the casting is solidified. The cooling time can be estimated from several thermodynamic properties of the metal, the maximum wall thickness of the casting, and the complexity of the die. A greater wall thickness will require a longer cooling time. The geometric complexity of the die also requires a longer cooling time because the additional resistance to the flow of heat.

4) Ejection. After the predetermined cooling time has passed, the die halves can be opened and an ejection mechanism can push the casting out of the die cavity. The time to open the die can be estimated from the dry cycle time of the machine and the ejection time is determined by the size of the casting's envelope and should include time for the casting to fall free of the die. The ejection mechanism must apply some force to eject the part because during cooling the part shrinks and adheres to the die. Once the casting is ejected, the die can be clamped shut for the next injection.

5) Trimming. During cooling, the material in the channels of the die will solidify attached to the casting. This excess material, along with any flash that has occurred, must be trimmed from the casting either manually via cutting or sawing, or using a trimming press. The time required to trim the excess material can be estimated from the size of the casting's envelope. The scrap material that results from this trimming is either discarded or can be reused in the die casting process. Recycled material may need to be reconditioned to the proper chemical composition before it can be combined with non-recycled metal and reused in the die casting process.

Words and terms

clamping　夹紧

injection　注射;注射剂

facilitate　帮助;促进,助长;使容易

lubrication　润滑,加油;油润

die halves　半模

sufficient force　足够压力

thermodynamic properties　热力性质

geometric complexity　几何复杂度

die cavity　模腔

shrinks　收缩

trimming　修剪

casting's envelope　铸件外壳

Questions

What are the steps of die casting?

Which is the first step in the process of clamping?

8.2　Cast metal

The main die casting alloys are: zinc, aluminium, magnesium, copper, lead, and tin; although uncommon, ferrous die casting is also possible. Specific die casting alloys include: zamak, zinc aluminium, aluminium and so on. The Aluminum Association (AA) standards: AA 380, AA 384, AA 386, AA 390; and AZ91D magnesium. The following is a summary of the advantages of each alloy:

Zinc: the easiest metal to cast; high ductility; high impact strength; easily plated; economical for small parts; promotes long die life.

Aluminium: lightweight; high dimensional stability for complex shapes and thin walls; good corrosion resistance; good mechanical properties; high thermal and electrical conductivity; retains strength at high temperatures.

Magnesium: the easiest metal to machine; excellent strength-to-weight ratio; lightest alloy commonly die cast.

Copper: high hardness; high corrosion resistance; highest mechanical properties of alloys die cast; excellent wear resistance; excellent dimensional stability; strength approaching that of steel parts.

Silicon tombac: high-strength alloy made of copper, zinc and silicon. Often used as an alternative for investment casted steel parts.

Lead and tin: high density; extremely close dimensional accuracy; used for special forms of corrosion resistance. Such alloys are not used in foodservice applications for public health reasons. Type metal, an alloy of lead, tin and antimony (with sometimes traces of copper) is used for casting hand-set type in letterpress printing and hot foil blocking. Traditionally cast in hand jerk moulds now predominantly die cast after the industrialisation of the type foundries. Around 1900 the slug casting machines came onto the market and added further automation, with sometimes dozens of casting machines at one newspaper office.

Maximum weight limits for aluminium, brass, magnesium and zinc castings are approximately 70 pounds (32 kg), 10 lb (4.5 kg), 44 lb (20 kg), and 75 lb (34 kg), respectively.

The material used defines the minimum section thickness and minimum draft required for a casting as outlined in the table below. The thickest section should be less than 13 mm (0.5 inch), but can be greater.

Words and terms

zinc　锌

lead　铅

tin　锡

ferrous die casting　铁类金属压模铸造

high ductility　高塑性

high impact strength　高冲击强度

corrosion resistance　耐腐蚀性

mechanical properties　力学性能

electrical conductivity　电导率

Questions

Drescribe briefly the main die casting alloys.

What are the characteristics of aluminum and magnesium alloy?

8.3　Advantages and disadvantages

1) Excellent dimensional accuracy (dependent on casting material, but typically 0.1 mm for the first 2.5 cm (0.005 inch for the first inch) and 0.02 mm for each additional centimeter (0.002 inch for each additional inch).

2) Smooth cast surfaces (Ra 1 mm to 2.5 mm).

3)Thinner walls can be cast as compared to sand and permanent mold casting (approximately 0.75 mm or 0.030 inch).

4)Inserts can be cast-in (such as threaded inserts, heating elements, and high strength bearing surfaces).

5)Reduces or eliminates secondary machining operations.

6)Rapid production rates.

7)Casting tensile strength as high as 415 megapascals (60 ksi).

8)Casting of low fluidity metals.

The main disadvantage to die casting is the very high capital cost. Both the casting equipment required and the dies and related components are very costly, as compared to most other casting processes. Therefore, to make die casting an economic process, a large production volume is needed. Other disadvantages are that the process is limited to high-fluidity metals, and casting weights must be between 30 g (1 oz) and 10 kg (20 lb). In the standard die casting process the final casting will have a small amount of porosity. This prevents any heat treating or welding, because the heat causes the gas in the pores to expand, which causes micro-cracks inside the part and exfoliation of the surface. Thus a related disadvantage of die casting is that it is only for parts in which softness is acceptable. Parts needing hardening (through hardening or case hardening) and tempering are not cast in dies.

Words and terms

dimensional accuracy　尺寸精度
casting material　浇注料
centimeter　厘米
approximately　大约
rapid production rates　快速生产率
low fluidity metals　低流动性金属
porosity　多孔性
welding　焊接性

Question

What are advantages and disadvantages of die casting?

9

Centrifugal casting

Centrifugal casting, sometimes called robocasting, is a metal casting process that uses centrifugal force to form cylindrical parts. This differs from most metal casting processes, which use gravity or pressure to fill the mold. In centrifugal casting, a permanent mold made from steel, cast iron, or graphite is typically used. However, the use of expendable sand molds is also possible. The casting process is usually performed on a horizontal centrifugal casting machine (vertical machines are also available) and includes the following steps, as shown in Fig. 9.1.

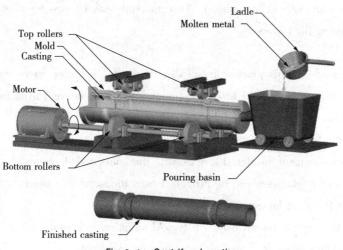

Fig. 9. 1　Centrifugal casting

1) Mold preparation. The walls of a cylindrical mold are first coated with a refractory ceramic coating, which involves a few steps (application, rotation, drying, and baking). Once prepared and secured, the mold is rotated about its axis at high speeds (300−3,000 rpm), typically around 1,000 rpm.

2) Pouring. Molten metal is poured directly into the rotating mold, without the use of runners or a gating system. The centrifugal force drives the material towards the mold walls as the mold fills.

3) Cooling. With all of the molten metal in the mold, the mold remains spinning as the metal cools. Cooling begins quickly at the mold walls and proceeds inwards.

4) Casting removal. After the casting has cooled and solidified, the rotation is stopped and the casting can be removed.

5) Finishing. While the centrifugal force drives the dense metal to the mold walls, any less dense impurities or bubbles flow to the inner surface of the casting. As a result, secondary processes such as machining, grinding, or sand-blasting, are required to clean and smooth the inner diameter of the part.

Centrifugal casting is used to produce axis-symmetric parts, such as cylinders or disks, which are typically hollow. Due to the high centrifugal forces, these parts have a very fine grain on the outer surface and possess mechanical properties approximately 30% greater than parts formed with static casting methods. These parts may be cast from ferrous metals such as low alloy steel, stainless steel, and iron, or from non-ferrous alloys such as aluminum, bronze, copper, magnesium, and nickel. Centrifugal casting is performed in wide variety of industries, including aerospace, industrial, marine, and power transmission. Typical parts include bearings, bushings, coils, cylinder liners, nozzles, pipes/tubes, pressure vessels, pulleys, rings, and wheels.

Materials

Typical materials that can be cast with this process are iron, steel, stainless steels, glass, and alloys of aluminum, copper and nickel. Two materials can be cast together by introducing a second material during the process.

Applications

Typical parts made by this process are pipes, flywheels, cylinder liners and other parts that are axisymmetric. It is notably used to cast cylinder liners and sleeve valves for piston engines, parts which could not be reliably manufactured otherwise.

Features of centrifugal casting

- Castings can be made in almost any length, thickness and diameter.
- Different wall thicknesses can be produced from the same size mold.
- Eliminates the need for cores.
- Resistant to atmospheric corrosion, a typical situation with pipes.
- Mechanical properties of centrifugal castings are excellent.
- Only cylindrical shapes can be produced with this process.
- Size limits are up to 6 m (20 feet) diameter and 15 m (50 feet) length.
- Wall thickness ranges from 2.5 mm to 125 mm (0.1 inch to 5.0 inch).
- Tolerance limit: on the OD can be 2.5 mm (0.1 inch), on the ID can be 3.8 mm (0.15 inch).
- Surface finish ranges from 2.5 mm to 12.5 mm (0.1 inch to 0.5 inch) rad/s.

Words and terms

centrifugal casting　离心浇铸造法

cylindrical parts　筒形件

graphite　石墨

ceramic coating　陶瓷涂层

bearings　轴承

bushings　衬套;套管;

coils　线圈

stainless steel　不锈钢

flywheel　调速轮

eliminate　消除;排除

Questions

Describe briefly the centrifugal casting.

What are the specific steps of centrifugal casting?

Show a brief overview of the characteristics of centrifugal casting.

10

Continuous casting

Continuous casting, also called strand casting, is the process whereby molten metal is solidified into a semifinished billet, bloom, or slab for subsequent rolling in the finishing mills. Prior to the introduction of continuous casting in the 1950s, steel was poured into stationary molds to form ingots. Since then, continuous casting has evolved to achieve improved yield, quality, productivity and cost efficiency. It allows lower-cost production of metal sections with better quality, due to the inherently lower costs of continuous, standardised production of a product, as well as providing increased control over the process through automation. This process is used most frequently to cast steel (in terms of tonnage cast). Aluminium and copper are also continuously cast.

10.1 Equipment and process

Molten metal is tapped into the ladle from furnaces. After undergoing any ladle treatments, such as alloying and degassing, and arriving at the correct temperature, the ladle is transported to the top of the casting machine, as shown in Fig 10.1. Usually the ladle sits in a slot on a rotating turret at the casting machine. One ladle is in the on-cast position (feeding the casting machine) while the other is made ready in the off-cast position, and is switched to the casting position when the first ladle is empty.

From the ladle, the hot metal is transferred via a refractory shroud (pipe) to a holding bath called a tundish. The tundish allows a reservoir of metal to feed the casting machine while ladles are switched, thus acting as a buffer of hot metal, as well as smoothing out flow, regulating metal feed to the molds and cleaning the metal.

Metal is drained from the tundish through another shroud into the top of an open-base copper mold. The depth of the mold can range from 0. 5 m to 2. 0 m (20 inch to 79 inch), depending on the casting speed and section size. The mold is water-cooled to solidify the hot metal directly in contact with it; this is the primary cooling process. It also oscillates vertically (or in a near vertical curved path) to prevent the metal sticking to the mold walls. A lubricant can also be added to the metal in the mold to prevent sticking, and to trap any slag particles—including oxide particles or scale—that may be present in the metal and bring them to the top of the pool to form a floating layer of slag. Often, the shroud is set so the hot metal exits it below the surface of the slag layer in the mold and is thus called a submerged entry nozzle (SEN). In some cases, shrouds may not be used between tundish and mold; in this case, interchangeable metering nozzles in the base of the tundish direct the metal into the moulds. Some continuous casting layouts feed several molds from the same tundish.

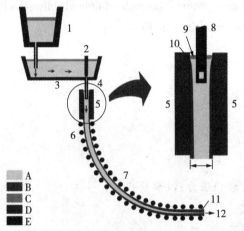

Notes: 1-Ladle. 2-Stopper. 3-Tundish. 4-Shroud. 5-Mold. 6-Roll support. 7-Turning zone. 8-Shroud. 9-Bath level. 10-Meniscus. 11-Withdrawal unit. 12-Slab.
A-Liquid metal. B-Solidified metal. C-Slag. D-Water-cooled copper plates. E-Refractory material.

Fig. 10. 1　Continuous casting

In the mold, a thin shell of metal next to the mold walls solidifies before the middle section, now called a strand, exits the base of the mold into a spray chamber. The bulk of metal within the walls of the strand is still molten. The strand is immediately supported by closely spaced, water-cooled rollers which support the walls of the strand against the ferrostatic pressure (compare hydrostatic pressure) of the still-solidifying liquid within the strand. To increase the rate of solidification, the strand is sprayed with large amounts of water as it passes through the spray-chamber; this is the secondary cooling process. Final solidification of the strand may take place after the strand has exited the spray-chamber.

It is here that the design of continuous casting machines may vary. This describes a curved apron casting machine; vertical configurations are also used. In a curved apron casting machine, the strand exits the mold vertically (or on a near vertical curved path) and as it travels through the spray-chamber, the rollers gradually curve the strand towards the horizontal. In a vertical casting machine, the strand stays vertical as it passes through the spray-chamber. Molds in a curved apron casting machine can be straight or curved, depending on the basic design of the machine.

In a true horizontal casting machine, the mold axis is horizontal and the flow of steel is horizontal from liquid to thin shell to solid (no bending). In this type of machine, either strand or mold oscillation is used to prevent sticking in the mold.

After exiting the spray-chamber, the strand passes through straightening rolls (if cast on other than a vertical machine) and withdrawal rolls. There may be a hot rolling stand after withdrawal to take advantage of the metal's hot condition to pre-shape the final strand. Finally, the strand is cut into predetermined lengths by mechanical shears or by travelling oxyacetylene torches, is marked for identification, and is taken either to a stockpile or to the next forming process.

In many cases the strand may continue through additional rollers and other mechanisms which may flatten, roll or extrude the metal into its final shape.

Words and terms

continuous casting　连续铸造
finishing mills　精轧机
ladle treatments　炉外处理
tundish　漏斗
ferrostatic pressure　钢水静压力;铁水静压力;
spray-chamber　喷雾室
reservoir　蓄水池;贮液器
submerged entry nozzle　浸入式水口

Questions

What is continuous casting?
What are the specific steps of continuous casting?
Show a brief overview of the characteristics of continuous casting.

10.2　Startup, control of the process and problems

Starting a continuous casting machine involves placing a dummy bar (essentially a curved metal beam) up through the spray chamber to close off the base of the mould. Metal is poured into the mould and withdrawn with the dummy bar once it solidifies. It is extremely important that the metal supply afterwards be guaranteed to avoid unnecessary shutdowns and restarts, known as turnarounds. Each time the caster stops and restarts, a new tundish is required, as any uncast metal in the tundish cannot be drained and instead freezes into a skull. Avoiding turnarounds requires the meltshop, including ladle furnaces (if any) to keep tight control on the temperature of the metal, which can vary dramatically with alloying additions, slag cover and deslagging, and the preheating of the ladle before it accepts metal, among other parameters. However, the cast rate may be lowered by reducing the amount of metal in the tundish (although this can increase wear on the tundish), or if the caster has multiple strands, one or more strands may be shut down to accommodate upstream delays. Turnarounds may be scheduled into a production sequence if the tundish temperature becomes too high after a certain number of heats or the service lifetime of a non-replaceable component (i. e., SEN in a thin-slab casting machine) is reached.

Many continuous casting operations are now fully computer-controlled. Several electromagnetic, thermal, or radiation sensors at the ladle shroud, tundish and mould sense the metal level or weight, flow rate and temperature of the hot metal, and the programmable logic controller (PLC) can set the rate of strand withdrawal via speed control of the withdrawal rolls. The flow of metal into the moulds can be controlled via two methods:

1) By slide gates or stopper rods at the top of the mould shrouds,

2) If the metal is open-poured, then the metal flow into the moulds is controlled solely by the internal diameter of the metering nozzles. These nozzles are usually interchangeable.

Overall casting speed can be adjusted by altering the amount of metal in the tundish, via the ladle slide gate. The PLC can also set the mould oscillation rate and the rate of mould powder feed, as well as the flow of water in the cooling sprays within the strand. Computer control also allows vital casting data to be transmitted to other manufacturing centers (particularly the steelmaking furnaces), allowing their work rates to be adjusted to avoid 'overflow' or 'underfund' of product.

Words and terms

startup 启动
dummy bar 引锭杆
shutdown 关机,关闭
deslagging 除渣
programmable logic controller(PLC) 可编程序逻辑控制器
withdrawal rolls 拉坯辊
metering nozzles 计量喷嘴

Questions

What problems can centrifugal casting solve?
What is the main role of PLC?

11

Principles of metal forming

11.1　Metal forming

Metal forming, is the metalworking process of fashioning metal parts and objects through mechanical deformation; the workpiece is reshaped without adding or removing material, and its mass remains unchanged. Forming operates on the materials science principle of plastic deformation, where the physical shape of a material is permanently deformed.

Metal forming tends to have more uniform characteristics across its subprocesses than its contemporary processes, cutting and joining. On the industrial scale, forming is characterized by: Very high loads and stresses required, between 50 N/mm^2 and 2,500 N/mm^2; Large, heavy, and expensive machinery in order to accommodate such high stresses and loads; Production runs with many parts, to maximize the economy of production and compensate for the expense of the machine tools.

In materials science, deformation refers to any changes in the shape or size of an object due to an applied force (the deformation energy in this case is transferred through work) or a change in temperature (the deformation energy in this case is transferred through heat). The first case can be a result of tensile (pulling) forces, compressive (pushing) forces, shear, bending or torsion (twisting).

In the second case, the most significant factor, which is determined by the temperature, is the mobility of the structural defects such as grain boundaries, point vacancies, line and screw dislocations, stacking faults and twins in both crystalline and non-crystalline solids. The movement or displacement of such mobile defects is thermally activated, and thus limited by the rate of atomic diffusion.

As deformation occurs, internal inter-molecular forces arise that oppose the applied force. If the applied force is not too great these forces may be sufficient to completely resist the applied force and allow the object to assume a new equilibrium state and to return to its original state when the load is removed. A larger applied force may lead to a permanent deformation of the object or even to its structural failure.

In Fig. 11.1 can be seen that the compressive loading (indicated by the arrow) has caused deformation in the cylinder so that the original shape (dashed lines) has changed (deformed) into one with bulging sides. The sides bulge because the material, although strong enough to not crack or otherwise fail, is not strong enough to support the load without change, thus the material is forced out laterally. Internal forces (in this case at right angles to the deformation) resist the applied load.

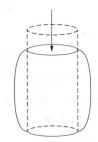

Fig. 11.1　Compressive stress results in deformation which shortens the object but also expands it outwards.

Depending on the type of material, size and geometry of the object, and the forces applied, various types of deformation may result. Different deformation modes may occur under different conditions, as can be depicted using a deformation mechanism map.

Words and terms

subprocesses　子步
mobility　流动能力;迁移率
point vacancy　空位
bulging　鼓起的
crack　裂纹
laterally　侧面地
geometry　几何形状

Questions

Do you know any material forming methods?
What are the factors that determine the material forming process?

11.2　Plastic deformation

This type of deformation is irreversible. However, an object in the plastic deformation range will first have undergone elastic deformation, which is reversible, so the object will return part way to its original shape. Soft thermoplastics have a rather large plastic deformation range as do ductile metals such as copper, silver, and gold. Steel does, too, but not cast iron. Hard thermosetting plastics, rubber, crystals, and ceramics have minimal plastic deformation ranges. An example of a material with a large plastic deformation range is wet chewing gum, which can be stretched dozens of times its original length.

Under tensile stress, plastic deformation is characterized by a strain hardening region and a necking region and finally, fracture (also called rupture). During strain hardening the material becomes stronger through the movement of atomic dislocations. The necking phase is indicated by a reduction in cross-sectional area of the specimen. Necking begins after the ultimate strength is reached. During necking, the material can no longer withstand the maximum stress and the strain in the specimen rapidly increases. Plastic deformation ends with the fracture of the material.

(1) Metal fatigue

Another deformation mechanism is metal fatigue, which occurs primarily in ductile metals. It was originally thought that a material deformed only within the elastic range returned completely to its original state once the forces were removed. However, faults are introduced at the molecular level with each deformation. After many deformations, cracks will begin to appear, followed soon after by a fracture, with no apparent plastic deformation in between. Depending on the material, shape, and how close to the elastic limit it is deformed, failure may require thousands, millions, billions, or trillions of deformations.

Metal fatigue has been a major cause of aircraft failure, especially before the process was well understood (for example, the De Havilland Comet accidents). There are two ways to determine when a part is in danger of metal fatigue; either predict when failure will occur due to the material/force/shape/iteration combination, and replace the vulnerable materials before this occurs, or perform inspections to detect the microscopic cracks and perform replacement once they occur. Selection of materials not likely to suffer from metal fatigue during the life of the product is the best solution, but not always possible. Avoiding shapes with sharp corners limits metal fatigue by reducing stress concentrations, but does not eliminate it.

Loading a structural element or specimen will increase the compressive stress until it reaches its compressive strength. According to the properties of the material, failure modes are yielding for materials with ductile behavior (most metals, some soils and plastics) or rupturing for brittle behavior (geomaterials, cast iron, glass, etc.).

In long, slender structural elements — such as columns or truss bars — an increase of compressive force F leads to structural failure due to buckling at lower stress than the compressive strength.

(2) Fracture

This type of deformation is also irreversible. A break occurs after the material has reached the end of the elastic, and then plastic, deformation ranges. At this point forces accumulate until they are sufficient to cause a fracture. All materials will eventually fracture, if sufficient forces are applied.

Words and terms

irreversible 无法复原的

thermoplastics 热塑性的

chewing gum 口香糖

strain hardening 应变强化

necking region 颈缩区

ultimate strength 极限强度

fatigue 疲劳

iteration 迭代

vulnerable 易受伤害的

Questions

What is the difference between elastic deformation and plastic deformation?

What are the characteristics of fatigue fracture?

11.3 Work hardening

Work hardening, also known as strain hardening or cold working, is the strengthening of a metal by plastic deformation. This strengthening occurs because of dislocation movements and dislocation generation within the crystal structure of the material. Many non-brittle metals with a reasonably high melting point as well as several polymers can be strengthened in this fashion. Alloys not amenable to heat treatment, including low-carbon steel, are often work-hardened. Some materials cannot be work-hardened at low temperatures, such as indium, however others can only be strengthened via work hardening, such as pure copper and aluminum, as shown in Fig. 11.2.

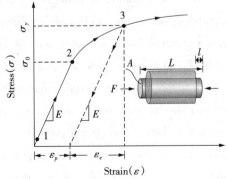

Fig. 11.2 A phenomenological uniaxial stress-strain curve showing typical work hardening plastic behavior of materials in uniaxial compression. For work hardening materials the yield stress increases with increasing plastic deformation.

Work hardening may be desirable or undesirable depending on the context. An example of undesirable work hardening is during machining when early passes of a cutter inadvertently work-harden the workpiece surface, causing damage to the cutter during the later passes. Certain alloys are more prone to this than others; superalloys such as Inconel require machining strategies that take it into account. An example of desirable work hardening is that which occurs in metal-working processes that intentionally induce plastic deformation to exact a shape change. These processes are known as cold working or cold forming processes. They are characterized by shaping the workpiece at a temperature below its recrystallization temperature, usually at ambient temperature. Cold forming techniques are usually classified into four major groups: squeezing, bending, drawing, and shearing. Applications include the heading of bolts and cap screws and the finishing of cold rolled steel. In cold forming, metal is formed at high speed and high pressure using tool steel or carbide dies. The cold working of the metal increases the hardness, yield strength, and tensile strength.

Before work hardening, the lattice of the material exhibits a regular, nearly defect-free pattern (almost no dislocations). The defect-free lattice can be created or restored at any time by annealing. As the material is work hardened it becomes increasingly saturated with new dislocations, and more dislocations are prevented from nucleating (a resistance to dislocation-formation develops). This resistance to dislocation-formation manifests itself as a resistance to plastic deformation; hence, the observed strengthening.

In metallic crystals, irreversible deformation is usually carried out on a microscopic scale by defects called dislocations, which are created by fluctuations in local stress fields within the material culminating in a lattice rearrangement as the dislocations propagate through the lattice. At normal temperatures the dislocations are not annihilated by annealing. Instead, the dislocations accumulate, interact with one another, and serve as pinning points or obstacles that significantly impede their motion. This leads to an increase in the yield strength of the material and a subsequent decrease in ductility.

Such deformation increases the concentration of dislocations which may subsequently form low-angle grain boundaries surrounding sub-grains. Cold working generally results in a higher yield strength as a result of the increased number of dislocations and the Hall-Petch effect of the sub-grains, and a decrease in ductility. The effects of cold working may be reversed by annealing the material at high temperatures where recovery and recrystallization reduce the dislocation density.

A material's work hardenability can be predicted by analyzing a stress-strain curve, or studied in context by performing hardness tests before and after a process.

Work hardening is a consequence of plastic deformation, a permanent change in shape. This is distinct from elastic deformation, which is reversible. Most materials do not exhibit only one or the other, but rather a combination of the two. The following discussion mostly applies to metals, especially steels, which are well studied. Work hardening occurs most notably for ductile materials such as metals. Ductility is the ability of a material to undergo plastic deformations before fracture (for example, bending a steel rod until it finally breaks).

The tensile test is widely used to study deformation mechanisms. This is because under compression, most materials will experience trivial (lattice mismatch) and non-trivial (buckling) events before plastic deformation or fracture occur. Hence the intermediate processes that occur to the material under uniaxial compression before the incidence of plastic deformation make the compressive test fraught with difficulties.

A material generally deforms elastically under the influence of small forces; the material returns quickly to its original shape when the deforming force is removed. This phenomenon is called elastic deformation. This behavior in materials is described by Hooke's Law. Materials behave elastically until the deforming force increases beyond the elastic limit, which is also known as the yield stress. At that point, the material is permanently deformed and fails to return to its original shape when the force is removed. This phenomenon is called plastic deformation. For example, if one stretches a coil spring up to a certain point, it will return to its original shape, but once it is stretched beyond the elastic limit, it will remain deformed and won't return to its original state.

Elastic deformation stretches the bonds between atoms away from their equilibrium radius of separation, without applying enough energy to break the inter-atomic bonds. Plastic deformation, on the other hand, breaks inter-atomic bonds, and therefore involves the rearrangement of atoms in a solid material.

In materials science parlance, dislocations are defined as line defects in a material's crystal structure. The bonds surrounding the dislocation are already elastically strained by the defect compared to the bonds between the constituents of the regular crystal lattice. Therefore, these bonds break at relatively lower stresses, leading to plastic deformation.

The strained bonds around a dislocation are characterized by lattice strain fields. For example, there are compressively strained bonds directly next to an edge dislocation and tensilely strained bonds beyond the end of an edge dislocation. These form compressive strain fields and tensile strain fields, respectively. Strain fields are analogous to electric fields in certain ways. Specifically, the strain fields of dislocations obey similar laws of attraction and repulsion; in order to reduce overall strain, compressive strains are attracted to tensile strains, and vice versa.

The visible (macroscopic) results of plastic deformation are the result of microscopic dislocation motion. For example, the stretching of a steel rod in a tensile tester is accommodated through dislocation motion on the atomic scale.

Increase in the number of dislocations is a quantification of work hardening. Plastic deformation occurs as a consequence of work being done on a material; energy is added to the material. In addition, the energy is almost always applied fast enough and in large enough magnitude to not only move existing dislocations, but also to produce a great number of new dislocations by jarring or working the material sufficiently enough. New dislocations are generated in proximity to a Frank-Read source.

Yield strength is increased in a cold-worked material. Using lattice strain fields, it can be shown that an environment filled with dislocations will hinder the movement of any one dislocation. Because dislocation motion is hindered, plastic deformation cannot occur at normal stresses. Upon application of stresses just beyond the yield strength of the non-cold-worked material, a cold-worked material will continue to deform using the only mechanism available: elastic deformation, the regular scheme of stretching or compressing of electrical bonds (without dislocation motion) continues to occur, and the modulus of elasticity is unchanged. Eventually the stress is great enough to overcome the strain-field interactions and plastic deformation resumes.

However, ductility of a work-hardened material is decreased. Ductility is the extent to which a material can undergo plastic deformation, that is, it is how far a material can be plastically deformed before fracture. A cold-worked material is, in effect, a normal (brittle) material that has already been extended through part of its allowed plastic deformation. If dislocation motion and plastic deformation have been hindered enough by dislocation accumulation, and stretching of electronic bonds and elastic deformation have reached their limit, a third mode of deformation occurs: fracture.

For an extreme example, in a tensile test a bar of steel is strained to just before the distance at which it usually fractures. The load is released smoothly and the material relieves some of its strain by decreasing in length. The decrease in length is called the elastic recovery, and the end result is a work-hardened steel bar. The fraction of length recovered (length recovered/original length) is equal to the yield-stress divided by the modulus of elasticity. (Here we discuss true stress in order to account for the drastic decrease in diameter in this tensile test.) The length recovered after removing a load from a material just before it breaks is equal to the length recovered after removing a load just before it enters plastic deformation.

The work-hardened steel bar has a large enough number of dislocations that the strain field interaction prevents all plastic deformation. Subsequent deformation requires a stress that varies linearly with the strain observed, the slope of the graph of stress vs. strain is the modulus of elasticity, as usual.

The work-hardened steel bar fractures when the applied stress exceeds the usual fracture stress and the strain exceeds usual fracture strain. This may be considered to be the elastic limit and the yield stress is now equal to the fracture toughness, which is of course, much higher than a non-work-hardened steel yield stress.

The amount of plastic deformation possible is zero, which is obviously less than the amount of plastic deformation possible for a non-work-hardened material. Thus, the ductility of the cold-worked bar is reduced.

Additionally, jewelers will construct structurally sound rings and other wearable objects (especially those worn on the hands) that require much more durability (than earrings for example) by utilizing a material's ability to be work hardened. While casting rings is done for a number of economical reasons (saving a great deal of time and cost of labor), a master jeweler may utilize the ability of a material to be work hardened and apply some combination of cold forming techniques during the production of a piece.

Words and terms

amenable　可用某种方式处理的

manifest　表明

fluctuation　波动

culminate　（在某一点）结束

annihilate　毁灭

pinning point　钉扎点

sub-grain　亚晶

equilibrium　平衡

Questions

What are the causes of work hardening?

Do you know any other ways to improve the strength of materials?

11.4　Recrystallization

Recrystallization is a process by which deformed grains are replaced by a new set of defects-free grains that nucleate and grow until the original grains have been entirely consumed. Recrystallization is usually accompanied by a reduction in the strength and hardness of a material and a simultaneous increase in the ductility. Thus, the process may be introduced as a deliberate step in metals processing or may be an undesirable byproduct of another processing step. The most important industrial uses are the softening of metals previously hardened by cold work, which have lost their ductility, and the control of the grain structure in the final product.

There are several, largely empirical laws of recrystallization:

1) Thermally activated. The rate of the microscopic mechanisms controlling the nucleation and growth of recrystallized grains depend on the annealing temperature. Arrhenius-type equations indicate an exponential relationship.

2) Critical temperature. Following from the previous rule it is found that recrystallization requires a minimum temperature for the necessary atomic mechanisms to occur. This recrystallization temperature decreases with annealing time.

3) Critical deformation. The prior deformation applied to the material must be adequate to provide nuclei and sufficient stored energy to drive their growth.

4) Deformation affects the critical temperature. Increasing the magnitude of prior deformation, or reducing the deformation temperature, will increase the stored energy and the number of potential nuclei. As a result, the recrystallization temperature will decrease with increasing deformation.

5) Initial grain size affects the critical temperature. Grain boundaries are good sites for nuclei to form. Since an increase in grain size results in fewer boundaries this results in a decrease in the nucleation rate and hence an increase in the recrystallization temperature.

6) Deformation affects the final grain size. Increasing the deformation, or reducing the deformation temperature, increases the rate of nucleation faster than it increases the rate of growth. As a result, the final grain size is reduced by increased deformation.

Words and terms

simultaneous 同时发生的
byproduct 副产物

Questions

What's the difference between recrystallization and crystallization?
Why increasing deformation can reduce recrystallization temperature?

11.5 Forming processes

Forming processes tend to be categorised by differences in effective stresses. These categories and descriptions are highly simplified, since the stresses operating at a local level in any given process are very complex and may involve many varieties of stresses operating simultaneously, or it may involve stresses which change over the course of the operation.

Compressive forming involves those processes where the primary means of plastic deformation is uni-or multiaxial compressive loading.

1) Rolling, where the material is passed through a pair of rollers,

2) Extrusion, where the material is pushed through an orifice,

3) Die forming, where the material is stamped by a press around or onto a die,

4) Forging, where the material is shaped by localized compressive forces,

5) Indenting, where a tool is pressed into the workpiece.

Tensile forming involves those processes where the primary means of plastic deformation is uni-or multiaxial tensile stress.

1) Stretching, where a tensile load is applied along the longitudinal axis of the workpiece,

2) Expanding, where the circumference of a hollow body is increased by tangential loading,

3) Recessing, where depressions and holes are formed through tensile loading.

Combined tensile and compressive forming. This category of forming processes involves those operations where the primary means of plastic deformation involves both tensile stresses and compressive loads.

1) Pulling through a die,

2) Deep drawing,

3) Spinning,

4) Flange forming,

5) Upset bulging,

6) Bending.

Words and terms

categorise　将……分类

description　说明

indenting　压痕

recessing　开槽

Question

How to distinguish tensile forming and compressive forming?

12 Forging

Forging is a manufacturing process involving the shaping of metal using localized compressive forces. The blows are delivered with a hammer (often a power hammer) or a die. Forging is often classified according to the temperature at which it is performed: cold forging (a type of cold working), warm forging, or hot forging (a type of hot working). For the latter two, the metal is heated, usually in a forge. Forged parts can range in weight from less than a kilogram to hundreds of metric tons. Forging has been done by smiths for millennia; the traditional products were kitchenware, hardware, hand tools, edged weapons, cymbals, and jewelry. Since the Industrial Revolution, forged parts are widely used in mechanisms and machines wherever a component requires high strength; such forgings usually require further processing (such as machining) to achieve a finished part. Today, forging is a major worldwide industry.

12. 1 Open-die forging

Open-die forging, also referred to as smith forging, blacksmith forging, hand forging, and flat-die forging, is generally performed without special tooling or only with simple tooling. The forgings obtained and the dimensions maintained are usually dependent upon the skill of the operator and the type of equipment used. However, with the addition of computer control to the equipment, more complex forgings can be produced and better dimensional control is maintained. Most open-die forgings are simple geometric shapes such as discs, rings, or shafts.

Open-die forging is also used in the steelmaking industry to cog ingot or to draw down billets from one size to a smaller one. The open-die forging process is employed when only a few part is needed and when the part is too large to be produced in closed dies. The open-die process is also used to obtain the mechanical properties in a workpiece that are not obtainable by machining. Generally most forgings begin with the open-die process before the final forging operation. Open-die forging is a hot forming process in which metal is shaped by hammering or pressing between flat or simple contoured dies.

- Upsetting, is a forging operation by which the length of a stock is reduced and the cross-section is increased (Fig. 12. 1).

- Drawing out, also referred to as stretching or swaging, is a forging operation by which a bar or slab is reduced in cross-section and increased in length (Fig. 12. 2).

- Punch expanding, is a forging operation by which the inner diameter of a hollow workblank is increased (Fig. 12. 3).

- Mandrel expanding, also known as saddle forging, involves continuous compressing a hollow workblank in turn along its circumference by means of the upper die and a mandrel supported by a saddle as shown in the figure. In mandrel expanding the workblank is not liable to break up because of little tangential tensile stress, so that this operation can be employed to produce thin-walled forging (Fig. 12. 4).

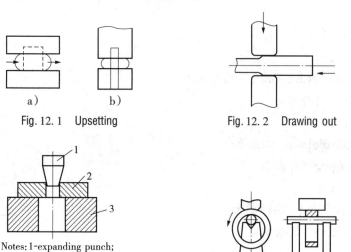

Fig. 12. 1　Upsetting

Fig. 12. 2　Drawing out

Notes: 1-expanding punch;
2-workblank; 3-pad ring

Fig. 12. 3　Punch expending

Fig. 12. 4　Mandrel expanding

12. 2　Close-die forging

- Close-die forging, also called impression-die forging, stamping, referred to the process in which the billet is formed by dies containing the impression of forging shape to be produced. The process is capable of production components of high quality at moderate cost. Forgings offer a high strength-to-weight ratio, toughness, and resistance to impact and fatigue.

- Close-die forging with flash (Fig. 12. 5)

　　In this process, a billet is formed (hot) in dies (usually with two halves) such that the flow of metal from the die cavity is restricted. The excess material is extruded through a restrictive narrow gap and appears as flash around the forging at the die parting line.

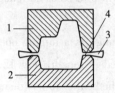

Notes: 1-upper die; 2-lower die; 3-flash; 4-flash land

Fig. 12. 5　Closed-die forging with flash.

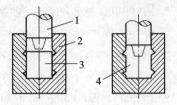

Notes: 1-punch; 2-split; 3-billet; 4-forging

Fig. 12. 6　Closed-die forging without flash

- Close-die forging without flash (Fig. 12. 6)

In this process, a billet with carefully controlled volume is deformed (hot or cold) by a punch in order to fill a die cavity without any loss of material. The punch and the die may be made up of one or several pieces.

Words and terms

forging　锻造
localized compressive forces　局部压力
kitchenware　厨房用具
jewelry　首饰
Industrial Revolution　工业革命
open-die forging　自由锻
upsetting　镦粗
mandrel expanding　芯棒扩大
close-die forging　闭模锻造

Questions

Briefly explain what forged is.

A simple description of the open-die forging process and close-die forging?

What is the biggest difference between open-die forging process and close-die forging?

13

Rolling

In metalworking, rolling is a metal forming process in which metal stock is passed through one or more pairs of rolls to reduce the thickness and to make the thickness uniform. The concept is similar to the rolling of dough. Rolling is classified according to the temperature of the metal rolled. If the temperature of the metal is above its recrystallization temperature, then the process is known as hot rolling. If the temperature of the metal is below its recrystallization temperature, the process is known as cold rolling. In terms of usage, hot rolling processes more tonnage than any other manufacturing process, and cold rolling processes the most tonnage out of all cold working processes. Roll stands holding pairs of rolls are grouped together into rolling mills that can quickly process metal, typically steel, into products such as bar steel (I-beams, angle stock, channel stock, and so on) and plate steel. Most steel mills have rolling mill divisions that convert the semi-finished casting products into finished products, as stown in Fig. 13.1.

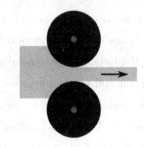

Fig. 13. 1   Rolling schematic view

There are many types of rolling processes, including ring rolling, roll bending, roll forming, profile rolling, and controlled rolling.

Words and terms

dough　生面团
tonnage　总重量
I-beams　工字梁
angle stock　角材
channel stock　槽材(U 型钢材)

Question

What are the roles of hot rolling and cold rolling?

13.1 **Hot rolling**

Hot rolling is a metalworking process that occurs above the recrystallization temperature of the material. After the grains deform during processing, they recrystallize, which maintains an equiaxed microstructure and prevents the metal from work hardening. The starting material is usually large pieces of metal, like semi-finished casting products, such as slabs, blooms, and billets. If these products came from a continuous casting operation, the products are usually fed directly into the rolling mills at the proper temperature. In smaller operations the material starts at room temperature and must be heated. This is done in a gas- or oil-fired soaking pit for larger workpieces and for smaller workpieces induction heating is used. As the material is worked the temperature must be monitored to make sure it remains above the recrystallization temperature. To maintain a safety factor a finishing temperature is defined above the recrystallization temperature; this is usually 50 ℃ to 100 ℃ above the recrystallization temperature. If the temperature does drop below this temperature the material must be re-heated before more hot rolling.

Hot rolled metals generally have little directionality in their mechanical properties and deformation induced residual stresses. However, in certain instances nonmetallic inclusions will impart some directionality and workpieces less than 20 mm thick often have some directional properties. Also, non-uniform cooling will induce a lot of residual stresses, which usually occurs in shapes that have a non-uniform cross-section, such as I-beams. While the finished product is of good quality, the surface is covered in mill scale, which is an oxide that forms at high temperatures. It is usually removed via pickling or the smooth clean surface process, which reveals a smooth surface. Dimensional tolerances are usually 2% to 5% of the overall dimension.

Hot rolled mild steel seems to have a wider tolerance for amount of included carbon than does cold rolled steel, and is therefore more difficult for a blacksmith to use. Also for similar metals, hot rolled products seem to be less costly than cold-rolled ones.

Hot rolling is used mainly to produce sheet metal or simple cross sections, such as rail tracks. Other typical uses for hot rolled metal includes truck frames, automotive wheels, pipe and tubular, water heaters, agriculture equipment, strappings, stampings, compressor shells, railcar components, wheel rims, metal buildings, railroad hopper cars, doors, shelving, discs, guard rails, automotive clutch plates.

Words and terms

slab　厚板
blooms　初轧坯
billet　钢坯
tolerance　公差
tubular　管子

Questions

Why does the metal need to be above its recrystallization temperature during hot rolling?

Why does hot rolling cost less than cold rolling?

13.2　Cold rolling

Cold rolling occurs with the metal below its recrystallization temperature (usually at room temperature), which increases the strength via strain hardening up to 20%. It also improves the surface finish and holds tighter tolerances. Commonly cold-rolled products include sheets, strips, bars, and rods; these products are usually smaller than the same products that are hot rolled. Because of the smaller size of the workpieces and their greater strength, as compared to hot rolled stock, four-high or cluster mills are used. Cold rolling cannot reduce the thickness of a workpiece as much as hot rolling in a single pass.

Cold-rolled sheets and strips come in various conditions: full-hard, half-hard, quarter-hard, and skin-rolled. Full-hard rolling reduces the thickness by 50%, while the others involve less of a reduction. Skin-rolling, also known as a skin-pass, involves the least amount of reduction: 0.5% to 1.0%. It is used to produce a smooth surface, a uniform thickness, and reduce the yield point phenomenon (by preventing Lüders bands from forming in later processing). It locks dislocations at the surface and thereby reduces the possibility of formation of Lüders bands. To avoid the formation of Lüders bands it is necessary to create substantial density of unpinned dislocations in ferrite matrix. It is also used to break up the spangles in galvanized steel. Skin-rolled stock is usually used in subsequent cold-working processes where good ductility is required.

Other shapes can be cold-rolled if the cross-section is relatively uniform and the transverse dimension is relatively small. Cold rolling shapes requires a series of shaping operations, usually along the lines of sizing, breakdown, roughing, semi-roughing, semi-finishing, and finishing.

If processed by a blacksmith, the smoother, more consistent, and lower levels of carbon encapsulated in the steel makes it easier to process, but at the cost of being more expensive.

Typical uses for cold-rolled steel include metal furniture, desks, filing cabinets, tables, chairs, motorcycle exhaust pipes, computer cabinets and hardware, home appliances and components, shelving, lighting fixtures, hinges, tubing, steel drums, lawn mowers, electronic cabinetry, water heaters, metal containers, fan blades, frying pans, wall and ceiling mount kits, and a variety of construction-related products.

Words and terms

Lüders bands　吕德斯带(屈服阶段出现的滑移带)

spangle　锌花

galvanized steel　镀锌钢

encapsulate　压缩

mower　剪草机

Questions

What are the advantages of cold rolling compared to hot rolling?

Why cold rolling plates are thinner than hot rolling ones?

13.3　Rolling processes

(1) Roll forming

Roll forming, roll bending or plate rolling is a continuous bending operation in which a long strip of metal (typically coiled steel) is passed through consecutive sets of rolls, or stands, each performing only an incremental part of the bend, until the desired cross-section profile is obtained. Roll forming is ideal for producing parts with long lengths or in large quantities. There are three main processes: 4 rollers, 3 rollers and 2 rollers, each of which has as different advantages according to the desired specifications of the output plate.

(2) Flat rolling

Flat rolling is the most basic form of rolling with the starting and ending material having a rectangular cross-section. The material is fed in between two rollers, called working rolls, that rotate in opposite directions. The gap between the two rolls is less than the thickness of the starting material, which causes it to deform. The decrease in material thickness causes the material to elongate. The friction at the interface between the material and the rolls causes the material to be pushed through. The amount of deformation possible in a single pass is limited by the friction between the rolls; if the change in thickness is too great the rolls just slip over the material and do not draw it in. The final product is either sheet or plate, with the former being less than 6 mm thick and the latter greater than; however, heavy plates tend to be formed using a press, which is termed forming, rather than rolling.

Often the rolls are heated to assist in the workability of the metal. Lubrication is often used to keep the workpiece from sticking to the rolls. To fine-tune the process, the speed of the rolls and the temperature of the rollers are adjusted.

The rolling is done in a cluster mill because the small thickness requires a small diameter rolls. To reduce the need for small rolls pack rolling is used, which rolls multiple sheets together to increase the effective starting thickness. As the foil sheets come through the rollers, they are trimmed and slitted with circular or razor-like knives. Trimming refers to the edges of the foil, while slitting involves cutting it into several sheets. Aluminum foil is the most commonly produced product via pack rolling. This is evident from the two different surface finishes; the shiny side is on the roll side and the dull side is against the other sheet of foil.

(3) Ring rolling

Ring rolling is a specialized type of hot rolling that increases the diameter of a ring. The starting material is a thick-walled ring. This workpiece is placed between two rolls, an inner idler roll and a driven roll, which presses the ring from the outside. As the rolling occurs the wall thickness decreases as the diameter increases. The rolls may be shaped to form various cross-sectional shapes. The resulting grain structure is circumferential, which gives better mechanical properties. Diameters can be as large as 8 m and face heights as tall as 2 m. Common applications include bearings, gears, rockets, turbines, airplanes, pipes, and pressure vessels.

(4) Controlled rolling

Controlled rolling is a type of thermomechanical processing which integrates controlled deformation and heat treating. The heat which brings the workpiece above the recrystallization temperature is also used to perform the heat treatments so that any subsequent heat treating is unnecessary. Types of heat treatments include the production of a fine grain structure; controlling the nature, size, and distribution of various transformation products (such as ferrite, austenite, pearlite, bainite, and martensite in steel); inducing precipitation hardening; and, controlling the toughness. In order to achieve this the entire process must be closely monitored and controlled. Common variables in controlled rolling include the starting material composition and structure, deformation levels, temperatures at various stages, and cool-down conditions. The benefits of controlled rolling include better mechanical properties and energy savings.

(5) Forge rolling

Forge rolling is a longitudinal rolling process to reduce the cross-sectional area of heated bars or billets by leading them between two contrary rotating roll segments. The process is mainly used to provide optimized material distribution for subsequent die forging processes. Owing to this a better material utilization, lower process forces and better surface quality of parts can be achieved in die forging processes.

Basically any forgeable metal can also be forge-rolled. Forge rolling is mainly used to preform long-scaled billets through targeted mass distribution for parts such as crankshafts, connection rods, steering knuckles and vehicle axles. Narrowest manufacturing tolerances can only partially be achieved by forge rolling. This is the main reason why forge rolling is rarely used for finishing, but mainly for preforming.

Characteristics of forge rolling: high productivity and high material utilization, good surface quality of forge-rolled workpieces, extended tool life-time, small tools and low tool costs, improved mechanical properties due to optimized grain flow compared to exclusively die forged workpieces.

(6) Profile rolling

Rolling process in which the workpiece passes through a series of rollers that gradually shapes the workpiece into the required profile.

The work piece [1], typically a raw bar, is first fed through a pair of rollers [2] to form a coarse beam. The beam is then passed on through multipurpose roller pairs [3] that have both horizontal and vertical rolls that are adjusted according to the desired profile. An edge roller [4] ensures that the upper and lower surfaces of the beam are parallel. The beam goes at last through another multipurpose roller pairs [5] which pushes out the final profile.

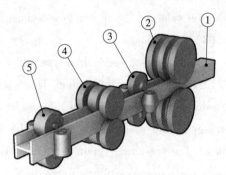

Fig. 13. 2   Profile rolling

Words and terms

incremental 增长的

lubrication 润滑

foil 箔

trim 修整

circumferential 周围的

utilization 利用

crankshafts 曲轴

steering knuckle 转向节

Question

What are the common characteristics of different rolling methods?

14

Extrusion

Extrusion is a process used to create objects of a fixed cross-sectional profile. A material is pushed through a die of the desired cross-section. The two main advantages of this process over other manufacturing processes are its ability to create very complex cross-sections, and to work materials that are brittle, because the material only encounters compressive and shear stresses. It also forms parts with an excellent surface finish.

Extrusion produces compressive and shear forces in the stock. No tensile is produced, which makes high deformation possible without tearing the metal, as shown in Fig. 14.1. The cavity in which the raw material is contained is lined with a wear resistant material. This can withstand the high radial loads that are created when the material is pushed the die.

Extrusions, often minimize the need for secondary machining, but are not of the same dimensional accuracy or surface finish as machined parts. Surface finish for steel is $3 \mu m$; and aluminum and magnesium is $0.8 \mu m$. However, this process can produce a wide variety of cross-sections that are hard to produce cost-effectively using other methods. Minimum thickness of steel is about 3 mm, whereas aluminum and magnesium is about 1 mm. Minimum cross sections are $250 \ mm^2$ for steel and less than that for aluminum and magnesium. Minimum corner and fillet radii are 0.4 mm for aluminum and magnesium, and for steel, the minimum corner radius is 0.8 mm and 4 mm fillet radius.

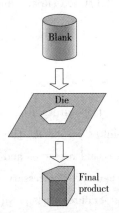

Fig. 14.1 Extrusion of a round blank through a die

Words and terms

extrusion 挤压

cross-sectional 截面

manufacture 制造

brittle 易脆的

shear stress 剪应力

raw material 原材料

resistant material 耐腐蚀材料

Questions

What is the extrusion process? How to classify?

What are the extrusion process parameters?

14.1　Cold extrusion

Cold extrusion is the process done at room temperature or slightly elevated temperatures. This process can be used for most materials-subject to designing robust enough tooling that can withstand the stresses created by extrusion. Examples of the metals that can be extruded are lead, tin, aluminum alloys, copper, titanium, molybdenum, vanadium, steel. Examples of parts that are cold extruded are collapsible tubes, aluminum cans, cylinders, gear blanks. The advantages of cold extrusion are:

1) No oxidation takes place;

2) Good mechanical properties due to severe cold working as long as the temperatures created are below the re-crystallization temperature;

3) Good surface finish with the use of proper lubricants.

14.2　Hot extrusion

Hot extrusion is done at fairly high temperatures, approximately 50% to 75% of the melting point of the metal. The pressures can range from 35 MPa to 700 MPa. Due to the high temperatures and pressures and its detrimental effect on the die life as well as other components, good lubrication is necessary. Oil and graphite work at lower temperatures, whereas at higher temperatures glass powder is used.

- Forward extrusion: In this process, a punch compresses a billet (hot or cold) confined in a container so that the billet material flows through a die in the same direction as the punch (Fig. 14.2).

- Backward extrusion: In this process, a moving punch applies a steady pressure to a slug (hot or cold) confined in a closed die, and forces the metal to flow around the punch in a direction opposite the direction of punch travel (Fig. 14.3).

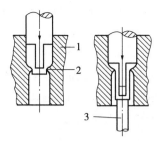

Notes: 1-die; 2-die bearing area;

3-ejector pin

Fig. 14. 2　Forward extrusion

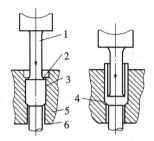

Notes: 1-punch; 2-bearing area; 3-billet;

4-workpiece; 5-die; 6-ejector pin

Fig. 14. 3　Backward extrusion

Extrusion may be continuous (theoretically producing indefinitely long material) or semi-continuous (producing many pieces). The extrusion process can be done with the material hot or cold. Commonly extruded materials include metals, polymers, ceramics, concrete, play dough, and foodstuffs.

14.3　**Materials**

(1) Metals that are commonly extruded include:

- Aluminium is the most commonly extruded material. Aluminium can be hot or cold extruded. If it is hot extruded it is heated to 575 ℉ to 1,100 ℉ (300 ℃ to 600 ℃). Examples of products include profiles for tracks, frames, rails, mullions, and heat sinks.

- Brass is used to extrude corrosion free rods, automobile parts, pipe fittings, engineering parts.

- Copper (1,100 ℉ to 1,825 ℉ (600 ℃ to 1,000 ℃)) pipe, wire, rods, bars, tubes, and welding electrodes. Often more than 690 MPa is required to extrude copper.

- Lead and tin (maximum 575 ℉ (300 ℃)) pipes, wire, tubes, and cable sheathing. Molten lead may also be used in place of billets on vertical extrusion presses.

- Magnesium (575 ℉ to 1,100 ℉ (300 ℃ to 600 ℃)) aircraft parts and nuclear industry parts. Magnesium is about as extrudable as aluminum.

- Zinc (400 ℉ to 650 ℉ (200 ℃ to 350 ℃)) rods, bar, tubes, hardware components, fitting, and handrails.

- Steel (1,825 ℉ to 2,375 ℉ (1,000 ℃ to 1,300 ℃)) rods and tracks. Usually plain carbon steel is extruded, but alloy steel and stainless steel can also be extruded.

- Titanium (1,100 ℉ to 1,825 ℉ (600 ℃ to 1,000 ℃)) aircraft components including seat tracks, engine rings, and other structural parts.

(2) Plastic

Plastic extrusion commonly uses plastic chips or pellets, which are usually dried, to drive out moisture, in a hopper before going to the feed screw. The polymer resin is heated to molten state by a combination of heating elements and shear heating from the extrusion screw.

（3）Ceramic

Ceramic can also be formed into shapes via extrusion. Terracotta extrusion is used to produce pipes. Many modern bricks are also manufactured using a brick extrusion process.

Words and terms

cold extrusion　冷挤压

hot extrusion　热挤压

forward extrusion　正挤压

backward extrusion　反挤压

elevated temperatures　高温

re-crystallization temperature　再结晶温度

good lubrication　良好的润滑

semi-continuous　半连续

polymer　多聚物；聚合物

plastic　塑性

ceramic　陶瓷

Questions

What is hot extrusion? What is cold extrusion? How to distinguish extrusion temperature?

Describe separately the hot extrusion process and the cold extrusion process.

What are the extruded metal materials? What should they pay attention to when they squeeze?

15
Drawing

Drawing is one of the oldest metal forming operations and has the major industrial significance. This process allows excellent surface finish and closely controlled dimensions to be obtained in long products that have constant cross section. In drawing, a previously rolled, extruded or fabricated product with solid or hollow cross sections is pulled through a die at a relatively high speed. In drawing of steel or aluminum wire, for example, exit speed of several thousand feet per minute are very common. The die geometry determines the final dimensions, the cross-sectional area of the drawn product, and the reduction area. Drawing is usually conducted at room temperature using a number of passes of reductions through consecutively located dies. At times, annealing may be necessary after a number of drawing passes before the drawing operation is continued. The deformation is accomplished by combination of tensile and compressive stresses that are created by the pulling force at the exit from the die, by the back-pull tensile force that is present between consecutive passes, and by the die configuration.

In wire or rod drawing, the section is usually round but could also be a shape. Fig. 15. 1 illustrates a few examples of drawn shapes. In cold drawing of shapes, the basic contour of the incoming shape is established by cold rolling passes that are usually preceded by annealing. After rolling, the section shape is refined and reduced to close tolerances by cold drawing. Here again, a number of annealing steps may be necessary to eliminate the effects of strain hardening, i. e., to reduce the flow stress and increase the ductility.

In tube drawing without a mandrel, also called tube sinking, the tube is initially pointed to facilitate feeding through the die; it is then reduced in outside diameter while the wall thickness and the tube length are increased.

Drawing with a fixed plug (Fig. 15. 2) is widely known and used for drawing of large-diameter to medium-diameter straight tubes. The plug, when pushed into the deformation zone, is pulled forward by the frictional force created by the sliding movement of the deforming tube. Thus, it is necessary to hold the plug in the correct position with a plug bar. In drawing of long and small diameter tubes, the plug bar may stretch and even break. In such cases it is advantageous to use a floating plug (Fig. 15. 3). This process can be used to draw any length of tubing by coiling the drawn tube at high speeds up to 2,000 ft/min (1 ft = 0. 304,8 m). In drawing with a moving mandrel, the mandrel travels at the speed at which the section exits from the die. This process, also called ironing, is widely used for thinning of the walls of drawn cups or shells, for example, in the production of beverage cans or artillery shells.

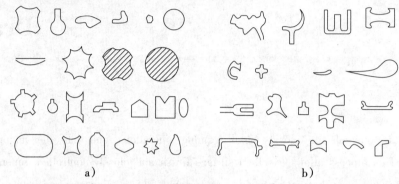

a)   b)

Fig. 15. 1   Example of cold drawn steel shapes drawn from round or square section
(a) and drawn from pre-rolled section (b)

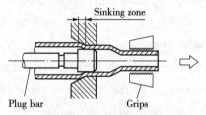

Fig. 15. 2   Schematic illustration of
drawing with a fixed plug

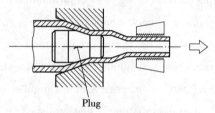

Fig. 15. 3   Schematic illustration of
drawing with a floating plug

Wire and rod drawing are very similar in principle. Normally, both are cold forming processes that produce excellent surface finish and close diameter tolerances, both start with hot rolled bar or rod. Hot rolled bar or rod is first freed from scale and dirt by pickling; it is then lubricated by coat with lime. Lime acts as a lubricant carrier and neutralizes any acid remaining from pickling. Rods that cannot be coiled are drawn in draw benches, at draw speeds of up to 500 ft/min. Wire can be coiled and therefore it is drawn through successive dies and rolled on bull blocks between drawing passes.

◇ **The cold drawing process for steel bars and wire**

- Coating: The surface of the bar or coil is coated with a drawing lubricant to aid cold drawing.

- Pointing: Several inches of the lead ends of the bar or coil are reduced in size by swaging or extruding so that it can pass freely through the drawing die. Note: This is done because the die opening is always smaller than the original bar or coil section size.

- Cold drawing: In this process, the material being drawn is at room temperature

(i. e. Cold-drawn). The pointed/reduced end of the bar or coil, which is smaller than the die opening, is passed through the die where it enters a gripping device of the drawing machine. The drawing machine pulls or draws the remaining unreduced section of the bar or coil through the die. The die reduces the cross section of the original bar or coil, shapes the profile of the product and increases the length of the original product.

- Finished Product: The drawn product, which is referred to as cold drawn or cold finished, exhibits a bright and/or polished finish, increased mechanical properties, improved machining characteristics and precise and uniform dimensional tolerances.

- Multi-pass drawing: The cold drawing of complex shapes/profiles may require that each bar/coil be drawn several times in order to produce the desired shape and tolerances. This process is called multi-pass drawing and involves drawing through smaller and smaller die openings. Material is generally annealed between each drawing pass to remove cold work and to increase ductility.

- Annealing: This is a thermal treatment generally used to soften the material being drawn, to modify the microstructure, the mechanical properties and the machining characteristics of the steel and/or to remove internal stresses in the product. Depending on the desired characteristics of the finished product, annealing may be used before, during (between passes) or after the cold drawing operation, depending on material requirements.

Words and terms

drawing　拉拔
die geometry　口模结构
tensile　拉伸
back-pull　后拉力
annealing　退火
tube sinking　缩口
beverage cans　饮料罐
artillery shells　炮弹
coating　涂层
tolerance　公差
multi-pass drawing　多道次拉拔

Questions

What is the drawing process?

What are the characteristics of the drawing process? What products can be prepared?

What are the main steps in cold drawing?

16

Stamping

Stamping, here, is used as a general term to cover all press working operations on sheet metal; it is not confined to forming and drawing processes. The stamping of parts from sheet metal is a straightforward operation, in which the metal is shaped or cut through deformation by shearing, punching, drawing, stretching, bending, coining, etc. Production rates are high and secondary machining is generally not required to produce finished parts within tolerances. A stamped part may be produced by one or a combination of three fundamental press operations (cutting, drawing and coining) applied to a given material.

(1) Blanking (Fig. 16. 1)

Blanking is an operation of cutting or shearing a piece out of a stock to a predetermined contour. The part (blank) that has been cut out is used, and the remaining portion of the original sheet or strip of metal becomes scrap. It is performed in a press with a die.

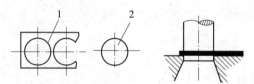

Notes: 1-scrap skeleton; 2-blank

Fig. 16. 1　Blanking

(2) Punching (Fig. 16. 2)

Punching involves the cutting of clean holes with resulting scrap slugs. Although the punching process is identical to that of blanking, the difference between the two operations lies in which part is used and which part is scrapped. In blanking, the part cut from the original metal is saved; in punching, the part that is cut out (called a slag) is scrapped and the original metal is saved.

Punching is performed in a die operated by a press. Piecing consists of cutting (or tearing) a hole in metal; however, it does not generate a slug. Instead, the metal is pushed back to form a jagged flange on the backside of the hole. A pierced hole looks somewhat like a bullet hole in a sheet of metal. Piercing is often confused with punching, in spite of the fact that the distinction between punching and piercing is clear, people are used the term "piercing" instead of "punching" sometimes.

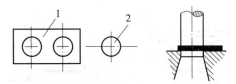

Notes: 1-blank; 2-scrap slug

Fig. 16. 2　Punching

（3）Shearing（Fig. 16. 3）

Shearing is the cutting action along a straight line to separate metal by two moving blades. The blades can be connected directly to a machine called a squaring shear or connected to a press by means of die sets. In shearing, a narrow strip of metal is plastically deformed to the point where it fractures at the surfaces in contact with the blades. The fracture operation propagates inward to provide separation. A chip does not form in this operation. The operation is used for producing blanks.

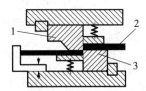

Notes: 1-moving blade; 2-metal; 3-stationary blade

Fig. 16. 3　Shearing

（4）Shaving（Fig. 16. 4）

Shaving involves cutting off metal in a chip fashion to obtain accurate dimension and also to remove the rough fractured edge of the sheet metal. Shaving is performed using dies with a very small clearance, as shown in the Fig. 16. 4. It is considered to be a secondary shearing operation.

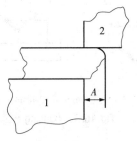

Notes: 1-die; 2-punch

Fig. 16. 4　Shaving

(5) Trimming (Fig. 16.5)

Trimming is a secondary cutting or shearing operation on previously formed, drawn parts in which the surplus metal or irregular outline or edge is sheared off to form desired shape and size.

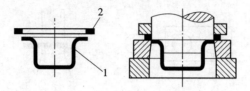

Notes: 1-cup after trimming; 2-scrap

Fig. 16.5　Trimming

(6) Lancing (Fig. 16.6)

Lancing is a slitting and forming process to produce a pocket-shaped opening in the product without freeing the scrap from the product. It is performed using a progressive die operation on a press. Lacing cuts are necessary to create louvers, which are formed in sheet metal for venting function.

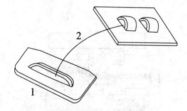

Notes: 1-louver; 2-lanced and formed, one operation

Fig. 16.6　Lancing

(7) Bending

Bending is a process by which a straight length of sheet metal is plastically deformed to a required curved length, as illustrated in Fig. 16.7. It is common forming operation for changing sheet and plate into channel, drums, tanks, etc.

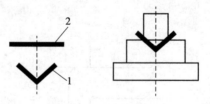

Notes: 1-workpiece; 2-blank

Fig. 16.7　Bending

(8) Flanging (Fig. 16.8)

Flanging is a bending operation in which the linen of bending is straight. During flanging, in the length direction there is no stresses imposed on the material; therefore, within reason, a flange of any desired width can be made.

Fig. 16. 8　Flanging

Words and terms

stamping　冲压

stretching　拉伸

blanking　下料

piercing　刺耳

punching　冲压,打孔

shaving　修面

flanging　卷边

Questions

What is the stamping process?

What are the characteristics of the stamping process?

What are the characteristics of each step in stamping process?

17

History of welding process

The history of joining metals goes back several millennia. Called forge welding, the earliest examples come from the Bronze and Iron Ages in Europe and the Middle East. The ancient Greek historian Herodotus states in The Histories of the 5th century BC that Glaucus of Chios "was the man who single-handedly invented iron welding". Welding was used in the construction of the Iron pillar of Delhi, erected in Delhi, India about 310 AD and weighing 5.4 metric tons.

The middle ages brought advances in forge welding, in which blacksmiths pounded heated metal repeatedly until bonding occurred. In 1540, Vannoccio Biringuccio published De la pirotechnia, which includes descriptions of the forging operation. Renaissance craftsmen were skilled in the process, and the industry continued to grow during the following centuries.

In 1800, Sir Humphry Davy discovered the short-pulse electrical arc and presented his results in 1801. In 1802, Russian scientist Vasily Petrov created the continuous electric arc, and subsequently published "News of Galvanic-Voltaic Experiments" in 1803, in which he described experiments carried out in 1802. Of great importance in this work was the description of a stable arc discharge and the indication of its possible use for many applications, one being melting metals. In 1808, Davy, who was unaware of Petrov's work, rediscovered the continuous electric arc. In 1881—1882 inventors Nikolai Benardos (Russian) and Stanislaw Olszewski (Polish) created the first electric arc welding method known as carbon arc welding using carbon electrodes. The advances in arc welding continued with the invention of metal electrodes in the late 1800s by a Russian, Nikolai Slavyanov (1888), and an American, C. L. Coffin (1890). Around 1900, A. P. Strohmenger released a coated metal electrode in Britain, which gave a more stable arc. In 1905, Russian scientist Vladimir Mitkevich proposed using a three-phase electric arc for welding. In 1919, alternating current welding was invented by C. J. Holslag but did not become popular for another decade.

Resistance welding was also developed during the final decades of the 19th century, with the first patents going to Elihu Thomson in 1885, who produced further advances over the next 15 years. Thermite welding was invented in 1893, and around that time another process, oxyfuel welding, became well established. Acetylene was discovered in 1836 by Edmund Davy, but its use was not practical in welding until about 1900, when a suitable torch was developed. At first, oxyfuel welding was one of the more popular welding methods due to its portability and relatively low cost. As the 20th century progressed, however, it fell out of favor for industrial applications. It was largely replaced with arc welding, as metal coverings (known as flux) for the electrode that stabilize the arc and shield the base material from impurities continued to be developed.

World War I caused a major surge in the use of welding processes, with the various military powers attempting to determine which of the several new welding processes would be best. The British primarily used arc welding, even constructing a ship, the "Fullagar" with an entirely welded hull. Arc welding was first applied to aircraft during the war as well, as some German airplane fuselages were constructed using the process. Also noteworthy is the first welded road bridge in the world, the Maurzyce Bridge designed by Stefan Bryla of the Lwów University of Technology in 1927, and built across the river Studwia near Lowicz towicz, Poland in 1928.

During the 1920s, major advances were made in welding technology, including the introduction of automatic welding in 1920, in which electrode wire was fed continuously. Shielding gas became a subject receiving much attention, as scientists attempted to protect welds from the effects of oxygen and nitrogen in the atmosphere. Porosity and brittleness were the primary problems, and the solutions that developed included the use of hydrogen, argon, and helium as welding atmospheres. During the following decade, further advances allowed for the welding of reactive metals like aluminum and magnesium. This in conjunction with developments in automatic welding, alternating current, and fluxes fed a major expansion of arc welding during the 1930s and then during World War II. In 1930, the first all-welded merchant vessel, M/S Carolinian, was launched.

During the middle of the century, many new welding methods were invented. In 1930, Kyle Taylor was responsible for the release of stud welding, which soon became popular in shipbuilding and construction. Submerged arc welding was invented the same year and continues to be popular today. In 1932 a Russian, Konstantin Khrenov successfully implemented the first underwater electric arc welding. Gas tungsten arc welding, after decades of development, was finally perfected in 1941, and gas metal arc welding followed in 1948, allowing for fast welding of non-ferrous materials but requiring expensive shielding gases. Shielded metal arc welding was developed during the 1950s, using a flux-coated consumable electrode, and it quickly became the most popular metal arc welding process. In 1957, the flux-cored arc welding process debuted, in which the self-shielded wire electrode could be used with automatic equipment, resulting in greatly increased welding speeds, and that same year, plasma arc welding was invented. Electroslag welding was introduced in 1958, and it was followed by its cousin, electrogas welding, in 1961. In 1953, the Soviet scientist N. F. Kazakov proposed the diffusion bonding method.

Other recent developments in welding include the 1958 breakthrough of electron beam welding, making deep and narrow welding possible through the concentrated heat source. Following the invention of the laser in 1960, laser beam welding debuted several decades later, and has proved to be especially useful in high-speed, automated welding. Magnetic pulse welding (MPW) is industrially used since 1967. Friction stir welding was invented in 1991 by Wayne Thomas at The Welding Institute (TWI, UK) and found high-quality applications all over the world. All of these four new processes continue to be quite expensive due the high cost of the necessary equipment, and this has limited their applications.

Words and terms

forge welding 锻焊

erected in *n.* 竖立在

electric arc *n.* 电弧

carry out 实施,贯彻

indication *n.* 指出;迹象;象征

hull 船体

fuselages 机身

hydrogen 氢气

helium 氦气

aluminum 铝

Bronze and Iron Age 青铜铁器时代

skilled in 熟悉,擅长于,擅长

bonding *n.* 粘合;*v.* 结合;*adj.* 结合的;

discharge 解雇;卸下;排放;解雇

oxyfuel welding 气焊

diffusion bonding 扩散焊

automatic welding 自动焊接

argon 氩,氩气

atmosphere 空气

magnesium 镁

conjunction 结合

expansion 膨胀

merchant vessel 商船

debut 初次登台

self-shielded wire electrode 自保护焊丝

alternating current 交流电

launch 发射,发动,使……下水

stud welding　电栓焊

non-ferrous material　有色金属材料

Questions

What's the earliest welding technique?

What's the main use of stud welding?

Among the welding techniques mentioned above, which have good portability and cost relatively low?

18
Oxyfuel gas welding and cutting

Oxyfuel welding (OFW) is a group of welding processes which join metals by heating them with a fuel gas flame or flares with or without the application of pressure and with or without the use of filler metal (Fig. 18. 1).

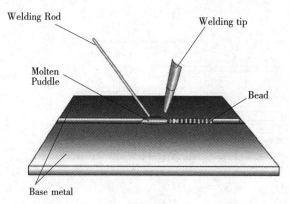

Fig. 18. 1   Diagram of oxyfuel gas welding

OFW includes any welding operation that makes use of a fuel gas combined with oxygen as a heating medium. The process involves the melting of the base metal and a filler metal, if used, by means of the flame produced at the tip of a welding torch.

Fuel gas and oxygen are mixed in the proper proportions in a mixing chamber which may be part of the welding tip assembly. Molten metal from the plate edges and filler metal, if used, intermix in a common molten pool. Upon cooling, they coalesce to form a continuous piece.

There are three major processes within the OFW group:

1) Oxyacetylene welding,

2) Oxyhydrogen welding,

3) Pressure gas welding.

There is one process of minor OFW, known as air acetylene welding, in which heat is obtained from the combustion of acetylene with air. Welding with methyl acetone-propadiene gas (MAPP gas) is also an oxyfuel procedure.

18.1 Oxyfuel gases

Commercial fuel gases have one common property: they all require oxygen to support combustion. To be suitable for welding operations, a fuel gas, when burned with oxygen, must have the following:

1) High flame temperature,

2) High rate of flame propagation,

3) Adequate heat content,

4) Minimum chemical reaction of the flame with base and filler metals.

There are three key advantages of OFW welding:

1) One advantage of this welding process is the control a welder can exercise over the rate of heat input, the temperature of the weld zone, and the oxidizing or reducing potential of the welding atmosphere.

2) Weld bead size and shape and weld puddle viscosity are also controlled in the welding process because the filler metal is added independently of the welding heat source.

3) OFW is ideally suited to the welding of thin sheet, tubes, and small diameter pipe. It is also used for repair welding. Thick section welds, except for repair work, are not economical.

Among the many reasons why welders use oxy-acetylene gas flames (OFW) is the high operating temperatures (5,589 ℉) and the ability to melt many common metals. Among the commercially available OFW, Oxy Fuel welding fuel gases, acetylene (combination of hydrogen and carbon) most closely meets all these requirements. Other gases, fuel such as MAPP gas, propylene, propane, natural gas, and proprietary gases based on these, have sufficiently high flame temperatures but exhibit low flame propagation rates.

These OFW, Oxy Fuel welding gas flames are excessively oxidizing at oxygen-to-fuel gas ratios high enough to produce usable heat transfer rates. Flame holding devices, such as counter bores on the tips, are necessary for stable operation and good heat transfer, even at the higher ratios. These OFW, Oxy Fuel welding gases, however, are used for oxygen cutting. They are also used for torch brazing, soldering, and many other operations where the demands upon the flame characteristics and heat transfer rates are not the same as those for welding.

Words and terms

flame 火焰

flare 闪光,燃烧,张开

combustion 燃烧,氧化

heat transfer　热传递

degrees Fahrenheit　华氏温度

spark　火花

heat content　焓,热含量

weld puddle viscosity　熔池黏度

torch brazing　火焰钎焊

Questions

Why is the oxyfuel gas welding so popular?

What's the main composition of oxyfuel gas?

List three applications of oxyfuel gas welding.

18.2　Types of flames

Three types of flames can be created with oxy-acetylene (OFW). Each one can be differentiated by the color, size and shape. Each flame can be created with a unique mixture of acetylene and oxygen (Fig. 18.2).

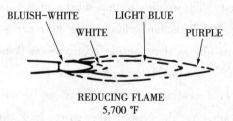

REDUCING FLAME
5,700 °F

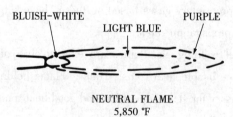

NEUTRAL FLAME
5,850 °F

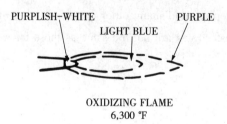

OXIDIZING FLAME
6,300 °F

Fig. 18.2　Typical flames of oxyfuel welding

1）Carburizing uses a mix of more acetylene compared to oxygen. Used for soldering, welding, braze welding and silver brazing.

2）Neutral is the most common type. It is created with a balanced mix of oxygen and acetylene. Causes a slow melt. It creates fewer sparks and does not boil. It protects steel from oxidation and the joint does not burn. It results in strong welds.

3）Oxidizing, as the name implies, uses a richer mix of oxygen than acetelyne. This type of flame isn't considered to be practical for welding. It can cause issues such as hardening at the weld, weld weakness and brittleness.

18.3 Oxyfuel welding fluxes

Oxides of all ordinary commercial metals have higher melting points than the metals and alloys (except steel) have themselves. They are usually pasty when the metal is quite fluid and at the proper welding temperature. An efficient flux will combine with oxides to form a fusible slag. The slag will have a melting point lower than the metal so it will flow away from the immediate field of action. It combines with base metal oxides and removes them. It also maintains cleanliness of the base metal at the welding area and helps remove oxide film on the surface of the metal. The welding area should be cleaned by any method. The flux also serves as a protection for the molten metal against atmospheric oxidation.

The chemical characteristics and melting points of the oxides of different metals vary greatly. There is no one flux that is satisfactory for all metals, and there is no national standard for gas welding fluxes. They are categorized according to the basic ingredient in the flux or base metal for which they are to be used.

Fluxes are usually in powder form. These fluxes are often applied by sticking the hot filler metal rod in the flux. Sufficient flux will adhere to the rod to provide proper fluxing action as the filler rod is melted in the flame.

Other types of fluxes are of a paste consistency which is usually painted on the filler rod or on the work to be welded.

Welding rods with a covering of flux are also available. Fluxes are available from welding supply companies and should be used in accordance with the directions accompanying them.

The melting point of a flux must be lower than that of either the metal or the oxides formed, so that it will be liquid. The ideal flux has exactly the right fluidity when the welding temperature has been reached. The flux will protect the molten metal from atmospheric oxidation. Such a flux will remain close to the weld area instead of flowing all over the base metal for some distance from the weld.

Fluxes differ in their composition according to the metals with which they are to be used. In cast iron welding, a slag forms on the surface of the puddle. The flux serves to break this up. Equal parts of a carbonate of soda and bicarbonate of soda make a good compound for this purpose. Nonferrous metals usually require a flux. Copper also requires a filler rod containing enough phosphorous to produce a metal free from oxides. Borax which has been melted and powdered is often used as a flux with copper alloys. A good flux is required with aluminum, because there is a tendency for the heavy slag formed to mix with the melted aluminum and weaken the weld. For sheet aluminum welding, it is customary to dissolve the flux in water and apply it to the rod. After welding aluminum, all traces of the flux should be removed.

Words and terms

carburizing　渗碳
boil　沸腾
contaminant　污染
groove　凹槽
ingredient　组成成分
compound　化合物
bicarbonate　碳酸氢盐
neutral　中立的
dirt　污垢
permit　许可
consumption　消耗,消费
powder　粉末
carbonate　碳酸盐
trace　痕迹

Questions

Why is it necessary to clean the surface of the workpiece before oxyfuel gas welding?
What's the main function of oxyfuel welding fluxes?

18.4　Cutting

If iron or steel is heated to its kindling temperature [not less than 1,600 °F (871 °C)], and is then brought into contact with oxygen, it burns or oxidizes very rapidly. The reaction of oxygen with the iron or steel forms iron oxide (Fe_3O_4) and gives off considerable heat. This heat is sufficient to melt the oxide and some of the base metal; consequently, more of the metal is exposed to the oxygen stream. This reaction of oxygen and iron is used in the oxyacetylene cutting process. A stream of oxygen is firmly fixed onto the metal surface after it has been heated to the kindling temperature. The hot metal reacts with oxygen, generating more heat and melting. The molten metal and oxide are swept away by the rapidly moving stream of oxygen. The oxidation reaction continues and furnishes heat for melting another layer of metal. The cut progresses in this manner.

Theoretically, the heat created by the burning iron would be sufficient to heat adjacent iron red hot, so that once started the cut could be continued indefinitely with oxygen only, as is done with the oxygen lance. In practice, however, excessive heat absorption at the surface caused by dirt, scale, or other substances, makes it necessary to keep the preheating flames of the torch burning throughout the operation.

19

Electric arc welding

※※※

19.1　Principle of arc welding

Arc welding is a process that is used to join metal to metal by using electricity to create enough heat to melt metal, and the melted metals when cool result in a binding of the metals. It is a type of welding that uses a welding power supply to create an electric arc between an electrode and the base material to melt the metals at the welding point. They can use either direct current (DC) or alternating current (AC), and consumable or non-consumable electrodes. The welding region is usually protected by some type of shielding gas, vapor, or slag. Arc welding processes may be manual, semi-automatic, or fully automated. First developed in the late part of the 19th century, arc welding became commercially important in shipbuilding during the World War II. Today it remains an important process for the fabrication of steel structures and vehicles.

19.2　Carbon arc welding

Carbon arc welding (CAW) is a process which produces coalescence of metals by heating them with an arc between a nonconsumable carbon (graphite) electrode and the work-piece. It was the first arc-welding process ever developed but is not used for many applications today, having been replaced by twin-carbon-arc welding and other variations. The purpose of arc welding is to form a bond between separate metals. In carbon-arc welding a carbon electrode is used to produce an electric arc between the electrode and the materials being bonded. This arc produces extreme temperatures in excess of 3,000 ℃. At this temperature the separate metals form a bond and become welded together, as shown in Fig. 19.1.

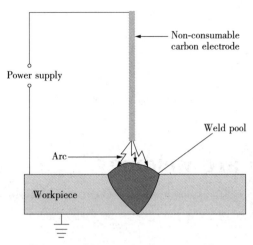

Fig. 19. 1   Diagram of carbon arc welding

CAW could not have been created if not for the discovery of the electric arc by Sir Humphry Davy in 1800, later repeated independently by a Russian physicist Vasily Vladimirovich Petrov in 1802. Petrov studied the electric arc and proposed its possible uses, including welding.

The inventors of carbon-arc welding were Nikolay Benardos and Stanistaw Olszewski, who developed this method in 1881 and patented it later under the name Elektrogefest.

Words and terms

electricity　电流

vapor　蒸汽

fabrication　制造

consumable　可消耗的

manual　手工的

extreme　极端的

patent　专利,获得专利

propose　建议,打算,计划

Questions

Which welding method produces a higher maximum temperature, the carbon arc welding or the oxyfuel welding?

Why is the carbon arc welding method replaced by other welding method? And what's the disadvantage of carbon arc welding?

20

Gas tungsten arc welding

Gas tungsten arc welding (GTAW), also known as tungsten inert gas (TIG) welding, is an arc welding process that uses a non-consumable tungsten electrode to produce the weld. The weld area is protected from atmospheric contamination by an inert shielding gas (argon or helium), and a filler metal is normally used, though some welds, known as autogenous welds, do not require it. A constant-current welding power supply produces electrical energy, which is conducted across the arc through a column of highly ionized gas and metal vapors known as plasma.

GTAW is most commonly used to weld thin sections of stainless steel and non-ferrous metals such as aluminum, magnesium, and copper alloys. The process grants the operator greater control over the weld than competing processes such as shielded metal arc welding and gas metal arc welding, allowing for stronger, higher quality welds. However, GTAW is comparatively more complex and difficult to master, and furthermore, it is significantly slower than most other welding techniques. A related process, plasma arc welding, uses a slightly different welding torch to create a more focused welding arc and as a result is often automated.

Fig. 20. 1 Diagram of TIG welding

20. 1 Applications

While the aerospace industry is one of the primary users of gas tungsten arc welding, the process is used in a number of other areas. Many industries use GTAW for welding thin workpieces, especially nonferrous metals. It is used extensively in the manufacture of space vehicles, and is also frequently employed to weld small-diameter, thin-wall tubing such as those used in the bicycle industry. In addition, GTAW is often used to make root or first-pass welds for piping of various sizes. In maintenance and repair work, the process is commonly used to repair tools and dies, especially components made of aluminum and magnesium. Because the weld metal is not transferred directly across the electric arc like most open arc welding processes, a vast assortment of welding filler metal is available to the welding engineer. In fact, no other welding process permits the welding of so many alloys in so many product configurations. Filler metal alloys, such as elemental aluminum and chromium, can be lost through the electric arc from volatilization. This loss does not occur with the GTAW process. Because the resulting welds have the same chemical integrity as the original base metal or match the base metals more closely, GTAW welds are highly resistant to corrosion and cracking over long time periods, making GTAW the welding procedure of choice for critical operations like sealing spent nuclear fuel canisters before burial.

Words and terms

dross-filled weld　渣填充焊

plasma　等离子

stabilize　使稳定

primary　主要的,初级的

volatilization　蒸发,挥发

autogenous　自发的,自生的

polarity　极性

maintenance　保持,维护

vehicle　车辆,交通工具

canister　桶,小罐

Questions

What's the greatest advantage of GTAW?

List some common applications of GTAW in daily life.

20.2　Quality

Gas tungsten arc welding, because it affords greater control over the weld area than other welding processes, can produce high-quality welds when performed by skilled operators. Maximum weld quality is assured by maintaining cleanliness—all equipments and materials used must be free from oil, moisture, dirt and other impurities, as these cause weld porosity and consequently a decrease in weld strength and quality. To remove oil and grease, alcohol or similar commercial solvents may be used, while a stainless steel wire brush or chemical process can remove oxides from the surfaces of metals like aluminum. Rust on steels can be removed by first grit blasting the surface and then using a wire brush to remove any embedded grit. These steps are especially important when negative polarity direct current is used, because such a power supply provides no cleaning during the welding process, unlike positive polarity direct current or alternating current. To maintain a clean weld pool during welding, the shielding gas flow should be sufficient and consistent so that the gas covers the weld and blocks impurities in the atmosphere. GTAW in windy or drafty environments increases the amount of shielding gas necessary to protect the weld, increasing the cost and making the process unpopular outdoors.

The level of heat input also affects weld quality. Low heat input, caused by low welding current or high welding speed, can limit penetration and cause the weld bead to lift away from the surface being welded. If there is too much heat input, however, the weld bead grows in width while the likelihood of excessive penetration and spatter increase. Additionally, if the welding torch is too far from the workpiece the shielding gas becomes ineffective, causing porosity within the weld. This results in a weld with pinholes, which is weaker than a typical weld.

If the amount of current used exceeds the capability of the electrode, tungsten inclusions in the weld may result. Known as tungsten spitting, this can be identified with radiography and can be prevented by changing the type of electrode or increasing the electrode diameter. In addition, if the electrode is not well protected by the gas shield or the operator accidentally allows it to contact the molten metal, it can become dirty or contaminated. This often causes the welding arc to become unstable, requiring that the electrode be ground with a diamond abrasive to remove the impurity.

20.3 Power supply

Gas tungsten arc welding uses a constant current power source, meaning that the current (and thus the heat) remains relatively constant, even if the arc distance and voltage change. This is important because most applications of GTAW are manual or semiautomatic, requiring that an operator hold the torch. Maintaining a suitably steady arc distance is difficult if a constant voltage power source is used instead, since it can cause dramatic heat variations and make welding more difficult.

The preferred polarity of the GTAW system depends largely on the type of metal being welded. Direct current with a negatively charged electrode (DCEN) is often employed when welding steels, nickel, titanium, and other metals. It can also be used in automatic GTAW of aluminum or magnesium when helium is used as a shielding gas. The negatively charged electrode generates heat by emitting electrons, which travel across the arc, causing thermal ionization of the shielding gas and increasing the temperature of the base material. The ionized shielding gas flows toward the electrode, not the base material, and this can allow oxides to build on the surface of the weld. Direct current with a positively charged electrode (DCEP) is less common, and is used primarily for shallow welds since less heat is generated in the base material. Instead of flowing from the electrode to the base material, as in DCEN, electrons go the other direction, causing the electrode to reach very high temperatures. To help it maintain its shape and prevent softening, a larger electrode is often used. As the electrons flow toward the electrode, ionized shielding gas flows back toward the base material, cleaning the weld by removing oxides and other impurities and thereby improving its quality and appearance.

Alternating current, commonly used when welding aluminum and magnesium manually or semi-automatically, combines the two direct currents by making the electrode and base material alternate between positive and negative charge. This causes the electron flow to switch directions constantly, preventing the tungsten electrode from overheating while maintaining the heat in the base material. Surface oxides are still removed during the electrode-positive portion of the cycle and the base metal is heated more deeply during the electrode-negative portion of the cycle. Some power supplies enable operators to use an unbalanced alternating current wave by modifying the exact percentage of time that the current spends in each state of polarity, giving them more control over the amount of heat and cleaning action supplied by the power source. In addition, operators must be wary of rectification, in which the arc fails to reignite as it passes from straight polarity (negative electrode) to reverse polarity (positive electrode). To remedy the problem, a square wave power supply can be used, as can high-frequency voltage to encourage ignition.

20.4　Electrode

The electrode used in GTAW is made of tungsten or a tungsten alloy, because tungsten has the highest melting temperature among pure metals, at 3,422 ℃ (6,192 ℉). As a result, the electrode is not consumed during welding, though some erosion (called burn-off) can occur. Electrodes can have either a clean finish or a ground finish—clean finish electrodes have been chemically cleaned, while ground finish electrodes have been ground to a uniform size and have a polished surface, making them optimal for heat conduction. The diameter of the electrode can vary between 0.5mm and 6.4 mm (0.02 inch and 0.25 inch), and their length can range from 75 mm to 610 mm(3.0 inch to 24.0 inch).

A number of tungsten alloys have been standardized by the International Organization for Standardization and the American Welding Society in ISO 6848 and AWS A5.12, respectively, for use in GTAW electrodes, and are summarized in the adjacent table.

Pure tungsten electrodes (classified as WP or EWP) are general purpose and low cost electrodes. They have poor heat resistance and electron emission. They find limited use in AC welding of magnesium and aluminum.

Cerium oxide (or ceria) as an alloying element improves arc stability and ease of starting while decreasing burn-off. Cerium addition is not as effective as thorium but works well, and cerium is not radioactive.

An alloy of lanthanum oxide (or lanthana) has a similar effect as cerium, and is also not radioactive.

Thorium oxide (or thoria) alloy electrodes offer excellent arc performance and starting, making them popular general purpose electrodes. However, it is somewhat radioactive, making inhalation of thorium vapors and dust a health risk, and disposal an environmental risk.

Electrodes containing zirconium oxide (or zirconia) increase the current capacity while improving arc stability and starting and increasing electrode life.

Filler metals are also used in nearly all applications of GTAW, the major exception being the welding of thin materials. Filler metals are available with different diameters and are made of a variety of materials. In most cases, the filler metal in the form of a rod is added to the weld pool manually, but some applications call for an automatically fed filler metal, which often is stored on spools or coils.

Words and terms

radioactive 放射性的

porosity 孔隙率

rust 锈,生锈

diamond abrasive 金刚石磨料

ignition 燃烧

cerium 铈

moisture 水分,湿度

solvent 溶剂,解决方法

grit blasting 喷砂处理

dramatic 急剧的

erosion 侵蚀,腐蚀

thorium 钍

Questions

What are the main factors that affect the quality of welded joints?

What kind of current types should be selected when welding magnesium alloy?

20.5　**Shielding gas**

As with other welding processes such as gas metal arc welding, shielding gases are necessary in GTAW to protect the welding area from atmospheric gases such as nitrogen and oxygen, which can cause fusion defects, porosity, and weld metal embrittlement if they come in contact with the electrode, the arc, or the welding metal. The gas also transfers heat from the tungsten electrode to the metal, and it helps start and maintain a stable arc.

The selection of a shielding gas depends on several factors, including the type of material being welded, joint design, and desired final weld appearance. Argon is the most commonly used shielding gas for GTAW, since it helps prevent defects due to a varying arc length. When used with alternating current, argon shielding results in high weld quality and good appearance. Another common shielding gas, helium, is most often used to increase the weld penetration in a joint, to increase the welding speed, and to weld metals with high heat conductivity, such as copper and aluminum. A significant disadvantage is the difficulty of striking an arc with helium gas, and the decreased weld quality associated with a varying arc length.

Argon-helium mixtures are also frequently utilized in GTAW, since they can increase control of the heat input while maintaining the benefits of using argon. Normally, the mixtures are made with primarily helium (often about 75% or higher) and a balance of argon. These mixtures increase the speed and quality of the AC welding of aluminum, and also make it easier to strike an arc. Another shielding gas mixture, argon-hydrogen, is used in the mechanized welding of light gauge stainless steel, but because hydrogen can cause porosity, its uses are limited. Similarly, nitrogen can sometimes be added to argon to help stabilize the austenite in austenitic stainless steels and increase penetration when welding copper. Due to porosity problems in ferritic steels and limited benefits, however, it is not a popular shielding gas additive.

Advantages and disadvantages

Advantages:

- Works on almost all types of metals with higher melting points. Gas tungsten arc welding is the most popular method for welding aluminum stainless steels, and nickel-base alloys. It is generally not used for the very low melting metals such as solders, or lead, tin, or zinc alloys. It is especially useful for joining aluminum and magnesium which form refractory oxides, and also for the reactive metals like titanium and zirconium, which dissolve oxygen and nitrogen and become embrittled if exposed to air while melting.

- Pinpoint accuracy and control. The process provides more precise control of the weld than any other arc welding process, because the arc heat and filler metal are independently controlled.

- Good looking weld beads.

- For metals of varying thickness including very thin metals (amperage range of 5 to 800, which is the amount of electricity created by the welding machine). The gas tungsten arc welding process is very good for joining thin base metals because of excellent control of heat input.

- Creates strong joints. It produces top quality welds in almost all metals and alloys used by industry.

- Clean process with minimal amount of fumes, sparks, spatter and smoke.

- High level of visibility when working due to low levels of smoke. Visibility is excellent because no smoke or fumes are produced during welding, and there is no slag or spatter that must be cleaned between passes or on a completed weld.

- Minimal finishing required. In very critical service applications or for very expensive metals or parts, the materials should be carefully cleaned of surface dirt, grease, and oxides before welding.

- Works in any position.

- TIG welding also has reduced distortion in the weld joint because of the concentrated heat source.

- As in oxyacetylene welding, the heat source and the addition of filler metal can be separately controlled.

- Because the electrode is non-consumable, the process can be used to weld by fusion alone without the addition of filler metal.

Disadvantages：

- Brighter UV rays when compared to other welding processes.

- Slower process than consumable electrode arc welding processes.

- Takes practice.

- More expensive process overall. Expensive welding supplies (vs. other processes) because the arc travel speed and weld metal deposition rates are lower than with some other methods. Inert gases for shielding and tungsten electrode costs add to the total cost of welding compared to other processes.

  Argon and helium used for shielding the arc are relatively expensive. Equipment costs are greater than that for other processes, such as shielded metal arc welding, which require less precise controls.

- Not easily portable, best for a welding shop.

- Transfer of molten tungsten from the electrode to the weld causes contamination. The resulting tungsten inclusion is hard and brittle.

- Exposure of the hot filler rod to air using improper welding techniques causes weld metal contamination.

Words and terms

atmospheric gas　大气

strike　冲击,打击,穿透

austenitic stainless　奥氏体不锈钢

grease　油脂

exposure　暴露,曝光

precise control　精确控制

nitrogen　氮气

austenite　奥氏体

nickel　镍

distortion　变形

portable　手提的,便携的

Questions

What kind of welding defects will occur when the flow of shielding gas is insufficient?

Is it true that the greater the shielding gas flow, the better of the welding quality? Why?

21

Gas metal arc welding

21.1 Overview

Gas metal arc welding (GMAW), sometimes referred to by its subtypes metal inert gas (MIG) welding or metal active gas (MAG) welding, is a welding process in which an electric arc forms between a consumable wire electrode and the workpiece metal(s), which heats the workpiece metal(s), causing them to melt and join.

Fig. 21. 1　Diagram of MIG welding

Along with the wire electrode, a shielding gas feeds through the welding gun, which shields the process from contaminants in the air, as shown in Fig. 21. 1. The process can be semi-automatic or automatic. A constant voltage, direct current power source is most commonly used with GMAW, but constant current systems, as well as alternating current, can be used. There are four primary methods of metal transfer in GMAW, called globular, short-circuiting, spray, and pulsed-spray, each of which has distinct properties and corresponding advantages and limitations.

Originally developed for welding aluminium and other non-ferrous materials in the 1940s, GMAW was soon applied to steels because it provided faster welding time compared to other welding processes. The cost of inert gas limited its use in steels until several years later, when the use of semi-inert gases such as carbon dioxide became common. Further developments during the 1950s and 1960s gave the process more versatility and as a result, it became a highly used industrial process. Today, GMAW is the most common industrial welding process, preferred for its versatility, speed and the relative ease of adapting the process to robotic automation. Unlike welding processes that do not employ a shielding gas, such as shielded metal arc welding, it is rarely used outdoors or in other areas of air volatility. A related process, flux cored arc welding, often does not use a shielding gas, but instead employs an electrode wire that is hollow and filled with flux.

The principles of gas metal arc welding began to be understood in the early 19th century, after Humphry Davy discovered the short pulsed electric arcs in 1800. Vasily Petrov independently produced the continuous electric arc in 1802 (followed by Davy after 1808). It was not until the 1880s that the technology became developed with the aim of industrial usage. At first, carbon electrodes were used in carbon arc welding. By 1890, metal electrodes had been invented by Nikolay Slavyanov and C. L. Coffin. In 1920, an early predecessor of GMAW was invented by P. O. Nobel of General Electric. It used a bare electrode wire and direct current, and used arc voltage to regulate the feed rate. It did not use a shielding gas to protect the weld, as developments in welding atmospheres did not take place until later that decade. In 1926 another forerunner of GMAW was released, but it was not suitable for practical use.

In 1948, GMAW was developed by the Battelle Memorial Institute. It used a smaller diameter electrode and a constant voltage power source developed by H. E. Kennedy. It offered a high deposition rate, but the high cost of inert gases limited its use to non-ferrous materials and prevented cost savings. In 1953, the use of carbon dioxide as a welding atmosphere was developed, and it quickly gained popularity in GMAW, since it made welding steel more economical. In 1958 and 1959, the short-arc variation of GMAW was released, which increased welding versatility and made the welding of thin materials possible while relying on smaller electrode wires and more advanced power supplies. It quickly became the most popular GMAW variation.

The spray-arc transfer variation was developed in the early 1960s, when experimenters added small amounts of oxygen to inert gases. More recently, pulsed current has been applied, giving rise to a new method called the pulsed spray-arc variation.

GMAW is one of the most popular welding methods, especially in industrial environments. It is used extensively by the sheet metal industry and, by extension, the automobile industry. There, the method is often used for arc spot welding, thereby replacing riveting or resistance spot welding. It is also popular for automated welding, in which robots handle the workpieces and the welding gun to speed up the manufacturing process. GMAW can be difficult to perform well outdoors, since drafts can dissipate the shielding gas and allow contaminants into the weld; flux cored arc welding is better suited for outdoor use such as in construction. Likewise, GMAW's use of a shielding gas does not lend itself to underwater welding, which is more commonly performed via shielded metal arc welding, flux cored arc welding, or gas tungsten arc welding.

21.2　Welding gun and wire feed unit

As illustrated in Fig. 21.1, The typical GMAW welding gun has a number of key parts—a control switch, a contact tip, a power cable, a gas nozzle, an electrode conduit and liner, and a gas hose. The control switch, or trigger, when pressed by the operator, initiates the wire feed, electric power, and the shielding gas flow, causing an electric arc to be struck. The contact tip, normally made of copper and sometimes chemically treated to reduce spatter, is connected to the welding power source through the power cable and transmits the electrical energy to the electrode while directing it to the weld area. It must be firmly secured and properly sized, since it must allow the electrode to pass while maintaining electrical contact. On the way to the contact tip, the wire is protected and guided by the electrode conduit and liner, which help prevent buckling and maintain an uninterrupted wire feed. The gas nozzle directs the shielding gas evenly into the welding zone. Inconsistent flow may not adequately protect the weld area. Larger nozzles provide greater shielding gas flow, which is useful for high current welding operations that develop a larger molten weld pool. A gas hose from the tanks of shielding gas supplies the gas to the nozzle. Sometimes, a water hose is also built into the welding gun, cooling the gun in high heat operations.

The wire feed unit supplies the electrode to the work, driving it through the conduit and on to the contact tip. Most models provide the wire at a constant feed rate, but more advanced machines can vary the feed rate in response to the arc length and voltage. Some wire feeders can reach feed rates as high as 30.5 m/min (1,200 inch/min), but feed rates for semiautomatic GMAW typically range from 2 m/min to 10 m/min (75 inch/ min/ to 400 inch/min).

Words and terms

subtype　图表类型

pulsed electric arc　脉冲电弧

dissipate　驱散,使……消散

gas nozzle　气焊嘴

workpiece　工件

draft　气流,草稿,起草

spatter　飞溅

tank　水箱

water hose　水管

conduit　沟槽,管子,管道

21.3 Power supply

Most applications of gas metal arc welding use a constant voltage power supply. As a result, any change in arc length (which is directly related to voltage) results in a large change in heat input and current. A shorter arc length causes a much greater heat input, which makes the wire electrode melt more quickly and thereby restore the original arc length. This helps operators keep the arc length consistent even when manually welding with hand-held welding guns. To achieve a similar effect, sometimes a constant current power source is used in combination with an arc voltage-controlled wire feed unit. In this case, a change in arc length makes the wire feed rate adjust to maintain a relatively constant arc length. In rare circumstances, a constant current power source and a constant wire feed rate unit might be coupled, especially for the welding of metals with high thermal conductivities, such as aluminum. This grants the operator additional control over the heat input into the weld, but requires significant skill to perform successfully.

Alternating current is rarely used with GMAW; instead, direct current is employed and the electrode is generally positively charged. Since the anode tends to have a greater heat concentration, this results in faster melting of the feed wire, which increases weld penetration and welding speed. The polarity can be reversed only when special emissive-coated electrode wires are used, but since these are not popular, a negatively charged electrode is rarely employed.

21.4 Electrode

Electrode selection is based primarily on the composition of the metal being welded, the process variation being used, joint design and the material surface conditions. Electrode selection greatly influences the mechanical properties of the weld and is a key factor of weld quality. In general the finished weld metal should have mechanical properties similar to those of the base material with no defects such as discontinuities, entrained contaminants or porosity within the weld. To achieve these goals a wide variety of electrodes exist. All commercially available electrodes contain deoxidizing metals such as silicon, manganese, titanium and aluminum in small percentages to help prevent oxygen porosity. Some contain denitriding metals such as titanium and zirconium to avoid nitrogen porosity. Depending on the process variation and base material being welded the diameters of the electrodes used in GMAW typically range from 0. 7 mm to 2. 4 mm (0. 028 inch to 0. 095 inch) but can be as large as 4 mm (0. 16 inch). The smallest electrodes, generally up to 1. 14 mm (0. 045 inch) are associated with the short-circuiting metal transfer process, while the most common spray-transfer process mode electrodes are usually at least 0. 9 mm (0. 035 inch).

21.5 Shielding gas

Shielding gases are necessary for gas metal arc welding to protect the welding area from atmospheric gases such as nitrogen and oxygen, which can cause fusion defects, porosity, and weld metal embrittlement if they come in contact with the electrode, the arc, or the welding metal. This problem is common to all arc welding processes; for example, in the older Shielded-Metal Arc Welding process (SMAW), the electrode is coated with a solid flux which evolves a protective cloud of carbon dioxide when melted by the arc. In GMAW, however, the electrode wire does not have a flux coating, and a separate shielding gas is employed to protect the weld. This eliminates slag, the hard residue from the flux that builds up after welding and must be chipped off to reveal the completed weld.

The choice of a shielding gas depends on several factors, most importantly the type of material being welded and the process variation being used. Pure inert gases such as argon and helium are only used for nonferrous welding; with steel they do not provide adequate weld penetration (argon) or cause an erratic arc and encourage spatter (with helium). Pure carbon dioxide, on the other hand, allows for deep penetration welds but encourages oxide formation, which adversely affect the mechanical properties of the weld. Its low cost makes it an attractive choice, but because of the reactivity of the arc plasma, spatter is unavoidable and welding thin materials is difficult. As a result, argon and carbon dioxide are frequently mixed in a 75%/25% to 90%/10% mixture. Generally, in short circuit GMAW, higher carbon dioxide content increases the weld heat and energy when all other weld parameters (volts, current, electrode type and diameter) are held the same. As the carbon dioxide content increases over 20%, spray transfer GMAW becomes increasingly problematic, especially with smaller electrode diameters.

Argon is also commonly mixed with other gases, oxygen, helium, hydrogen and nitrogen. The addition of up to 5% oxygen (like the higher concentrations of carbon dioxide mentioned above) can be helpful in welding stainless steel, however, in most applications carbon dioxide is preferred. Increased oxygen makes the shielding gas oxidize the electrode, which can lead to porosity in the deposit if the electrode does not contain sufficient deoxidizers. Excessive oxygen, especially when used in application for which it is not prescribed, can lead to brittleness in the heat affected zone. Argon-helium mixtures are extremely inert, and can be used on nonferrous materials. A helium concentration of 50% to 75% raises the required voltage and increases the heat in the arc, due to helium's higher ionization temperature. Hydrogen is sometimes added to argon in small concentrations (up to about 5%) for welding nickel and thick stainless steel workpieces. In higher concentrations (up to 25% hydrogen), it may be used for welding conductive materials such as copper. However, it should not be used on steel, aluminum or magnesium because it can cause porosity and hydrogen embrittlement.

Shielding gas mixtures of three or more gases are also available. Mixtures of argon, carbon dioxide and oxygen are marketed for welding steels. Other mixtures add a small amount of helium to argon-oxygen combinations, these mixtures are claimed to allow higher arc voltages and welding speed. Helium also sometimes serves as the base gas, with small amounts of argon and carbon dioxide added. However, because it is less dense than air, helium is less effective at shielding the weld than argon—which is denser than air. It also can lead to arc stability and penetration issues, and increased spatter, due to its much more energetic arc plasma. Helium is also substantially more expensive than other shielding gases. Other specialized and often proprietary gas mixtures claim even greater benefits for specific applications.

The desirable rate of shielding-gas flow depends primarily on weld geometry, speed, current, the type of gas, and the metal transfer mode. Welding flat surfaces requires higher flow than welding grooved materials, since gas disperses more quickly. Faster welding speeds, in general, mean that more gas must be supplied to provide adequate coverage. Additionally, higher current requires greater flow, and generally, more helium is required to provide adequate coverage than if argon is used. Perhaps most importantly, the four primary variations of GMAW have differing shielding gas flow requirements—for the small weld pools of the short circuiting and pulsed spray modes, about 10 L/min ($20 \text{ ft}^3/\text{h}$) is generally suitable, whereas for globular transfer, around 15 L/min ($30 \text{ ft}^3/\text{h}$) is preferred. The spray transfer variation normally requires more shielding-gas flow because of its higher heat input and thus larger weld pool. Typical gas-flow amounts are approximately 20 L/min to 25 L/min ($40 \text{ ft}^3/\text{h}$ to $50 \text{ ft}^3/\text{h}$).

Words and terms

grant　授予,同意

heat concentration　热集中

feed wire　填丝

denitriding metal　脱氮

spray-transfer　喷射过渡

helium　氦气

deoxidizing　脱氧金属

zirconium　锆

chip off　切去

argon　氩,氩气

Questions

What's the difference between GTAW and GMAW in electrode?

What kind of shielding gases should be used when welding aluminum and carbon steel, respectively? Why?

21.6 **Quality**

Two of the most prevalent quality problems in GMAW are dross and porosity. If not controlled, they can lead to weaker, less ductile welds. Dross is an especially common problem in aluminium GMAW welds, normally coming from particles of aluminium oxide or aluminum nitride present in the electrode or base materials. Electrodes and workpieces must be brushed with a wire brush or chemically treated to remove oxides on the surface. Any oxygen in contact with the weld pool, whether from the atmosphere or the shielding gas, causes dross as well. As a result, sufficient flow of inert shielding gases is necessary, and welding in volatile air should be avoided.

In GMAW the primary cause of porosity is gas entrapment in the weld pool, which occurs when the metal solidifies before the gas escapes. The gas can come from impurities in the shielding gas or on the workpiece, as well as from an excessively long or violent arc. Generally, the amount of gas entrapped is directly related to the cooling rate of the weld pool. Because of its higher thermal conductivity, aluminum welds are especially susceptible to greater cooling rates and thus additional porosity. To reduce it, the workpiece and electrode should be clean, the welding speed diminished and the current set high enough to provide sufficient heat input and stable metal transfer but low enough that the arc remains steady. Preheating can also help reduce the cooling rate in some cases by reducing the temperature gradient between the weld area and the base material.

21.7 **Metal transfer modes**

The three transfer modes in GMAW are globular, short-circuiting, and spray. There are a few recognized variations of these three transfer modes including modified short-circuiting and pulsed-spray.

21.7.1 **Globular**

GMAW with globular metal transfer is considered the least desirable of the three major GMAW variations, because of its tendency to produce high heat, a poor weld surface, and spatter. The method was originally developed as a cost efficient way to weld steel using GMAW, because this variation uses carbon dioxide, a less expensive shielding gas than argon. Adding to its economic advantage was its high deposition rate, allowing welding speeds of up to 110 mm/s (250 inch/min). As the weld is made, a ball of molten metal from the electrode tends to build up on the end of the electrode, often in irregular shapes with a larger diameter than the electrode itself. When the droplet finally detaches either by gravity or short circuiting, it falls to the workpiece, leaving an uneven surface and often causing spatter. As a result of the large molten droplet, the process is generally limited to flat and horizontal welding positions, requires thicker workpieces, and results in a larger weld pool.

21.7.2　Short-circuiting

Further developments in welding steel with GMAW led to a variation known as short-circuit transfer (SCT) or short-arc GMAW, in which the current is lower than for the globular method. As a result of the lower current, the heat input for the short-arc variation is considerably reduced, making it possible to weld thinner materials while decreasing the amount of distortion and residual stress in the weld area. As in globular welding, molten droplets form on the tip of the electrode, but instead of dropping to the weld pool, they bridge the gap between the electrode and the weld pool as a result of the lower wire feed rate. This causes a short circuit and extinguishes the arc, but it is quickly reignited after the surface tension of the weld pool pulls the molten metal bead off the electrode tip. This process is repeated about 100 times per second, making the arc appear constant to the human eye. This type of metal transfer provides better weld quality and less spatter than the globular variation, and allows for welding in all positions, albeit with slower deposition of weld material. Setting the weld process parameters (volts, amps and wire feed rate) within a relatively narrow band is critical to maintaining a stable arc: generally between 100 A and 200 A at 17 V to 22 V for most applications. Also, using short-arc transfer can result in lack of fusion and insufficient penetration when welding thicker materials, due to the lower arc energy and rapidly freezing weld pool. Like the globular variation, it can only be used on ferrous metals.

21.7.3　Spray

Spray transfer GMAW was the first metal transfer method used in GMAW, and well-suited to welding aluminium and stainless steel while employing an inert shielding gas. In this GMAW process, the weld electrode metal is rapidly passed along the stable electric arc from the electrode to the workpiece, essentially eliminating spatter and resulting in a high-quality weld finish. As the current and voltage increases beyond the range of short circuit transfer the weld electrode metal transfer transitions from larger globules through small droplets to a vaporized stream at the highest energies. Since this vaporized spray transfer variation of the GMAW weld process requires higher voltage and current than short circuit transfer, and as a result of the higher heat input and larger weld pool area (for a given weld electrode diameter), it is generally used only on workpieces of thicknesses above about 6.4 mm (0.25 inch).

Also, because of the large weld pool, it is often limited to flat and horizontal welding positions and sometimes also used for vertical-down welds. It is generally not practical for root pass welds. When a smaller electrode is used in conjunction with lower heat input, its versatility increases. The maximum deposition rate for spray arc GMAW is relatively high—about 60 mm/s (150 inch/min).

21.7.4　Pulsed-spray

A variation of the spray transfer mode, pulse-spray is based on the principles of spray transfer but uses a pulsing current to melt the filler wire and allow one small molten droplet to fall with each pulse. The pulses allow the average current to be lower, decreasing the overall heat input and thereby decreasing the size of the weld pool and heat-affected zone while making it possible to weld thin workpieces. The pulse provides a stable arc and no spatter, since no short-circuiting takes place. This also makes the process suitable for nearly all metals, and thicker electrode wire can be used as well. The smaller weld pool gives the variation greater versatility, making it possible to weld in all positions. In comparison with short arc GMAW, this method has a somewhat slower maximum speed (85 mm/s or 200 inch/min) and the process also requires that the shielding gas be primarily argon with a low carbon dioxide concentration. Additionally, it requires a special power source capable of providing current pulses with a frequency between 30 and 400 pulses per second. However, the method has gained popularity, since it requires lower heat input and can be used to weld thin workpieces, as well as nonferrous materials.

Words and terms

prevalent　流行的,普遍的

nitride　氮化物

entrap　使陷入

deposition rate　沉积速率

detach　分离

short-circuiting　短路

spray　喷射

dross　渣滓,碎屑

violent arc　强烈的电弧

globular　球状的

droplet　小滴,微滴

gravity　重力

albeit　虽然,即便

versatility　多样性,多功能性

Questions

Which kind of transfer mode is most expected when welding nonferrous alloy? Why?

Which kind of transfer mode provides the fastest welding speed for GMAW? Why?

22

Plasma arc welding

Plasma arc welding(PAW) is an arc welding process wherein coalescence is produced by the heat obtained from a constricted arc setup between a tungsten/alloy tungsten electrode and the water-cooled (constricting) nozzle (non-transferred arc) or between a tungsten/alloy tungsten electrode and the job (transferred arc). The process employs two inert gases, one forms the arc plasma and the second shields the arc plasma. Filler metal may or may not be added. The structure of the welding equipment is shown in Fig. 22.1(a).

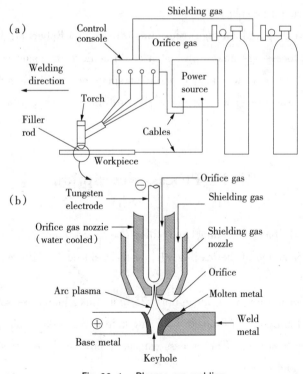

Fig. 22.1 Plasma arc welding

PAW is an arc welding process similar to gas tungsten arc welding (GTAW). The electric arc is formed between an electrode (which is usually but not always made of sintered tungsten) and the workpiece. The key difference from GTAW is that in PAW, by positioning the electrode within the body of the torch, the plasma arc can be separated from the shielding gas envelope. The plasma is then forced through a fine-bore copper nozzle which constricts the arc and the plasma exits the orifice at high velocities (approaching the speed of sound) and a temperature approaching 28,000 ℃ (50,000 ℉) or higher.

Just as oxy-fuel torches can be used for either welding or cutting, so too can plasma torches, which can achieve plasma arc welding or plasma cutting.

Arc plasma is the temporary state of a gas. The gas gets ionized after passage of electric current through it and it becomes a conductor of electricity. In ionized state atoms break into electrons (−) and cations (+) and the system contains a mixture of ions, electrons and highly excited atoms. The degree of ionization may be between 1% and greater than 100% i. e.; double and triple degrees of ionization. Such states exist as more electrons are pulled from their orbits.

The energy of the plasma jet and thus the temperature is dependent upon the electrical power employed to create arc plasma. A typical value of temperature obtained in a plasma jet torch may be of the order of 28,000 ℃ (50,000 ℉) against about 5,500 ℃ (10,000 ℉) in ordinary electric welding arc. Actually all welding arcs are (partially ionized) plasmas, but the one in plasma arc welding is a constricted arc plasma.

The plasma arc welding and cutting process was invented by Robert M. Gage in 1953 and patented in 1957. The process was unique in that it could achieve precision cutting and welding on both thin and thick metals. It was also capable of spray coating hardening metals onto other metals. One example was the spray coating of the turbine blades of the moon bound Saturn rocket.

22.1 Process description

Technique of work piece cleaning and filler metal addition is similar to that in TIG welding. Filler metal is added at the leading edge of the weld pool. Filler metal is not required in making root pass weld.

Type of Joints: For welding work piece up to 25 mm thick, joints like square butt, J or V are employed. Plasma welding is used to make both key hole and non-key hole types of welds.

Making a non-key hole weld: The process can make non key hole welds on work pieces having thickness 2. 4 mm and under.

Making a keyhole welds: An outstanding characteristics of plasma arc welding, owing to exceptional penetrating power of plasma jet, is its ability to produce keyhole welds in work piece having thickness from 2.5 mm to 25 mm. A keyhole effect is achieved through right selection of current, nozzle orifice diameter and travel speed, which create a forceful plasma jet to penetrate completely through the work piece. Plasma jet in no case should expel the molten metal from the joint. The major advantages of keyhole technique are the ability to penetrate rapidly through relatively thick root sections and to produce a uniform under bead without mechanical backing. Also, the ratio of the depth of penetration to the width of the weld is much higher, resulting narrower weld and heat-affected zone. As the weld progresses, base metal ahead the keyhole melts, flow around the same solidifies and forms the weld bead. Key holing aids deep penetration at faster speeds and produces high quality bead. While welding thicker pieces, in laying others than root run, and using filler metal, the force of plasma jet is reduced by suitably controlling the amount of orifice gas.

Plasma arc welding is an advancement over the GTAW process. This process uses a non-consumable tungsten electrode and an arc constricted through a fine-bore copper nozzle. PAW can be used to join all metals that are weldable with GTAW (i. e., most commercial metals and alloys). Difficult-to-weld in metals by PAW include bronze, cast iron, lead and magnesium. Several basic PAW process variations are possible by varying the current, plasma gas flow rate, and the orifice diameter, including:

- Micro-plasma (<15 A).
- Melt-in mode (15 A to 100 A).
- Keyhole mode (>100 A).
- Plasma arc welding has a greater energy concentration as compared to GTAW.
- A deep, narrow penetration is achievable, with a maximum depth of 12 mm to 18 mm (0.47 inch to 0.71 inch) depending on the material.
- Greater arc stability allows a much longer arc length (stand-off), and much greater tolerance to arc length changes.
- PAW requires relatively expensive and complex equipment as compared to GTAW; proper torch maintenance is critical.
- Welding procedures tend to be more complex and less tolerant to variations in fit-up, etc.
- Operator skill required is slightly greater than for GTAW.
- Orifice replacement is necessary.

Words and terms

overheat 过热

envelope 包层,信封

orifice 口孔

keyhole 锁眼

coalescence 合并,联合
fine-bore copper nozzle 细孔铜喷头
velocity 速度
plasma jet 等离子流

Question

What is Plasma arc welding?

22. 2 **Plasma torch**

The welding torch for plasma arc welding is similar in appearance to a gas tungsten arc torch, but more complex(Fig. 22. 2).

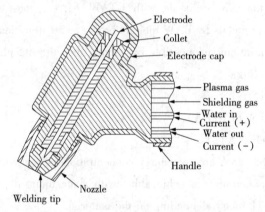

Fig. 22. 2   A cross section of a plasma welding arc torch head

All plasma torches are water cooled, even the lowest-current range torch. This is because the arc is contained inside a chamber in the torch where it generates considerable heat. If water flow is interrupted briefly, the nozzle may melt.

A cross section of a plasma welding arc torch head is shown by Fig. 22. 1(b). During the non-transferred period, the arc will be struck between the nozzle or tip with the orifice and the tungsten electrode. Manual plasma arc torches are made in various sizes starting with 100 A through 300 A. Automatic torches for machine operation are also available.

The torch utilizes the 2% thoriated tungsten electrode similar to that used for gas tungsten welding. Since the tungsten electrode is located inside the torch, it is almost impossible to contaminate it with base metal.

22.3　Wire feeder

A wire feeder may be used for machine or automatic welding and must be the constant speed type. The wire feeder must have a speed adjustment covering the range of from 10 inch/min (254 mm/min) to 125 inch/min (3.18 m/min) feed speed.

22.4　Shielding gases

Two inert gases or gas mixtures are employed. The orifice gas at lower pressure and flow rate forms the plasma arc. The pressure of the orifice gas is intentionally kept low to avoid weld metal turbulence, but this low pressure is not able to provide proper shielding of the weld pool. To have suitable shielding protection same or another inert gas is sent through the outer shielding ring of the torch at comparatively higher flow rates. Most of the materials can be welded with argon, helium, argon+hydrogen and argon+helium, as inert gases or gas mixtures. Argon is very commonly used. Helium is preferred where a broad heat input pattern and flatter cover pass is desired without key hole mode weld. A mixture of argon and hydrogen supplies heat energy higher than when only argon is used and thus permits keyhole mode welds in nickel base alloys, copper base alloys and stainless steels.

For cutting purposes a mixture of argon and hydrogen (10% to 30%) or that of nitrogen may be used. Hydrogen, because of its dissociation into atomic form and thereafter recombination generates temperatures above those attained by using argon or helium alone. In addition, hydrogen provides a reducing atmosphere, which helps in preventing oxidation of the weld and its vicinity. (Care must be taken, as hydrogen diffusing into the metal can lead to embrittlement in some metals and steels.)

At least two separate (and possibly three) flows of gas are used in PAW:
- Plasma gas flows through the orifice and becomes ionized.
- Shielding gas flows through the outer nozzle and shields the molten weld from the atmosphere.
- Back-purge and trailing gas required for certain materials and applications.

These gases can all be same, or of differing composition.

22.5　Key process variables

- Current Type and Polarity.

- DCEN from a CC source is standard.
- AC square-wave is common on aluminum and magnesium.
- Welding current and pulsing—current can vary from 0. 5 A to 1,200 A; Current can be constant or pulsed at frequencies up to 20 kHz.
- Gas flow rate (This critical variable must be carefully controlled based upon the current, orifice diameter and shape, gas mixture, and the base material and thickness).

22.6 Advantages and disadvantages

Advantages:

Advantages of plasma arc welding when compared to gas tungsten arc welding stem from the fact that PAW has a higher energy concentration. Its higher temperature, constricted cross-sectional area, and the velocity of the plasma jet create a higher heat content. The other advantage is based on the stiff columnar type of arc or form of the plasma, which doesn't flare like the gas tungsten arc.

These two factors provide the following advantages:

- More Freedom During Manual Welds: the torch-to-work distance from the plasma arc is less critical than for gas tungsten arc welding. This is important for manual operation, since it gives the welder more freedom to observe and control the weld.
- Keyhole Effect (complete single pass penetration): high temperature and high heat concentration of the plasma allow for the keyhole effect, which provides complete penetration single pass welding of many joints. In this operation, the heat affected zone and the form of the weld are more desirable. The heat-affected zone is smaller than with the gas tungsten arc, and the weld tends to have more parallel sides, which reduces angular distortion.

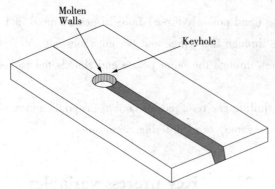

Fig. 22. 3 Keyhole mode

In the keyhole mode, a penetrating hole is formed at the leading edge of the weld pool. The molten weld metal flows around the hole and solidifies behind the keyhole to form the weld bead. Therefore, keyhole welds are complete penetration welds with high depth to width ratios. This results in low weld distortion. With operating currents up to 300 A, this mode can be used to weld materials up to about 3/4 inch thick, and to weld titanium and aluminum alloys.

- Faster Travel Speeds: the higher heat concentration and the plasma jet allow for higher travel speeds. The plasma arc is more stable and is not as easily deflected to the closest point of base metal. Greater variation in joint alignment is possible with plasma arc welding. This is important when making root pass welds on pipe and other one-side weld joints. Plasma welding has deeper penetration capabilities and produces a narrower weld. This means that the depth-to-width ratio is more advantageous.

Disadvantages:

- Orifice replacement necessary
- Expensive equipment
- More skill needed than for GTAW process

Words and terms

    chamber 室,膛

    control console 操作台

    delay timing system 延时系统

    torch protective device 电弧保护设备

    thoriate 涂钍

    pilot arc 维持电弧

    ammeter 电流计

    interlock 互锁

    dissociation 分离,分解

    back-purge 逆向清楚

    leading edge 前沿

    vicinity 邻近,附近

    angular distortion 角变形

Questions

What's the fundamental of TIG welding process?

What's in common between TIG and MIG welding? And what's difference between them?

What shielding gases are utilized in plasma welding?

23

Submerged arc welding

Submerged arc welding (SAW) is a common arc welding process. The first patent on the submerged-arc welding (SAW) process was taken out in 1935 and covered an electric arc beneath a bed of granulated flux. Originally developed and patented by Jones, Kennedy and Rothermund, the process requires a continuously fed consumable solid or tubular (metal cored) electrode. The molten weld and the arc zone are protected from atmospheric contamination by being "submerged" under a blanket of granular fusible flux consisting of lime, silica, manganese oxide, calcium fluoride, and other compounds. When molten, the flux becomes conductive, and provides a current path between the electrode and the work. This thick layer of flux completely covers the molten metal thus preventing spatter and sparks as well as suppressing the intense ultraviolet radiation and fumes that are a part of the shielded metal arc welding (SMAW) process.

The submerged arc welding process is shown by Fig. 23. 1. It utilizes the heat of an arc between a continuously fed electrode and the work.

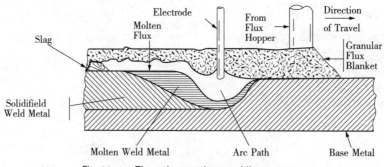

Fig. 23. 1　The submerged arc welding process

The heat of the arc melts the surface of the base metal and the end of the electrode. The metal melted off the electrode is transferred through the arc to the workpiece, where it becomes the deposited weld metal. Shielding is obtained from a blanket of granular flux, which is laid directly over the weld area. The flux close to the arc melts and intermixes with the molten weld metal, helping to purify and fortify it. The flux forms a glass-like slag that is lighter in weight than the deposited weld metal and floats on the surface as a protective cover. The weld is submerged under this layer of flux and slag, hence the name submerged arc welding. The flux and slag normally cover the arc so that it is not visible. The unmelted portion of the flux can be reused. The electrode is fed into the arc automatically from a coil. The arc is maintained automatically. Travel can be manual or by machine. The arc is initiated by a fuse type start or by a reversing or retrack system.

23.1　Key SAW process variables

- Wire feed speed (main factor in welding current control)
- Arc voltage
- Travel speed
- Electrode stick-out (ESO) or contact tip to work (CTTW)
- Polarity and current type (AC or DC) and variable balance AC current

In submerged arc welding, the electrode type and the flux type are usually based on the mechanical properties required by the weld. The electrode size is related to the weld joint size and the current recommended for the particular joint. This must also be considered in determining the number of passes or beads for a particular joint. Welds for the same joint dimension can be made in many or few passes, depending on the weld metal metallurgy desired. Multiple passes usually deposit higher-quality weld metal. Polarity is established initially and is based on whether maximum penetration or maximum deposition rate is required.

The major variables that affect the weld involve heat input and include the welding current, arc voltage, and travel speed. Welding current is the most important. For single-pass welds, the current should be sufficient for the desired penetration without burn-through. The higher the current, the deeper the penetration. In multi-pass work, the current should be suitable to produce the size of the weld expected in each pass. The welding current should be selected based on the electrode size. The higher the welding current, the greater the melt-off rate (deposition rate).

The arc voltage is varied within narrower limits than welding current. It has an influence on the bead width and shape.

Higher voltages will cause the bead to be wider and flatter. Extremely high arc voltage should be avoided, since it can cause cracking. This is because an abnormal amount of flux is melted and excess deoxidizers may be transferred to the weld deposit, lowering its ductility. Higher arc voltage also increases the amount of flux consumed. The low arc voltage produces a stiffer arc that improves penetration, particularly in the bottom of deep grooves. If the voltage is too low, a very narrow bead will result. It will have a high crown and the slag will be difficult to remove.

Travel speed influences both bead width and penetration. Faster travel speeds produce narrower beads that have less penetration. This can be an advantage for sheet metal welding where small beads and minimum penetration are required. If speeds are too fast, however, there is a tendency for undercut and porosity, since the weld freezes quicker. If the travel speed is too slow, the electrode stays in the weld puddle too long. This creates poor bead shape and may cause excessive spatter and flash through the layer of flux.

The secondary variables include the angle of the electrode to the work, the angle of the work itself, the thickness of the flux layer, and the distance between the current pickup tip and the arc. This latter factor, called electrode "stickout," has a considerable effect on the weld. Normally, the distance between the contact tip and the work is 1 inch to 1.5 inch. (25 mm to 38 mm). If the stickout is increased beyond this amount, it will cause preheating of the electrode wire, which will greatly increase the deposition rate. As stickout increases, the penetration into the base metal decreases. This factor must be given serious consideration because in some situations the penetration is required.

The depth of the flux layer must also be considered. If it is too thin, there will be too much arcing through the flux or arc flash. This also may cause porosity. If the flux depth is too heavy, the weld may be narrow and humped. Too many small particles in the flux can cause surface pitting since the gases generated in the weld may not be allowed to escape. These are sometimes called peck marks on the bead surface.

Words and terms

submerged arc welding　埋弧焊
granular　颗粒状的
calcium fluoride　氟化钙
fume　烟,冒烟
purify　净化,使纯净
establish　建立
stickout　杰出的,杰出人物
beneath　在……下方
manganese oxide　氧化锰
ultraviolet radiation　紫外线
fortify　加强,增强
fuse　使……熔化,保险丝
crown　王冠,加冕

Questions

What's the function of flux during submerged arc welding?

List three important parameters for controlling the heat input of submerged arc welding, and explain why.

23.2　Applications

The submerged arc process is widely used in heavy steel plate fabrication work. This includes the welding of structural shapes, the longitudinal seam of larger diameter pipe, the manufacture of machine components for all types of heavy industry, and the manufacture of vessels and tanks for pressure and storage use. It is widely used in the shipbuilding industry for splicing and fabricating sub-assemblies, and by many other industries where steels are used in medium to heavy thicknesses. It is also used for surfacing and buildup work, maintenance, and repair. Submerged arc welding is used to weld low- and medium-carbon steels, low-alloy high-strength steels, quenched and tempered steels, and many stainless steels.

Experimentally, it has been used to weld certain copper alloys, nickel alloys, and even uranium. Metal thicknesses from 1/16 inch to 1/2 inch (1.6 mm to 12.7 mm) can be welded with no edge preparation. With edge preparation, welds can be made with a single pass on material from 1/4 inch to 1 inch (6.4 mm to 25.4 mm).

When multipass technique is used, the maximum thickness is practically unlimited. Horizontal fillet welds can be made up to 3/8 inch (9.5 mm) in a single pass and in the flat position, fillet welds can be made up to 1 inch (25 mm) size.

23.3　Application restriction

The most popular method of SAW application is the machine method, where the operator monitors the welding operation.

Second in popularity is the automatic method, where welding is a pushbutton operation. The process can be applied semi-automatically; however, this method of application is not too popular.

The process cannot be applied manually because it is impossible for a welder to control an arc that is not visible. The submerged arc welding process is a limited-position welding process. The welding positions are limited because the large pool of molten metal and the slag are very fluid and will tend to run out of the joint. Welding can be done in the flat position and in the horizontal fillet position with ease. Under special controlled procedures, it is possible to weld in the horizontal position, sometimes called 3 o'clock welding. This requires special devices to hold the flux up so that the molten slag and weld metal cannot run away. The process cannot be used in the vertical or overhead position.

23.4　Welding circuit and current

The SAW or submerged arc welding process uses either direct or alternating current for welding power. Direct current is used for most applications which use a single arc. Both direct current electrode positive (DCEP) and electrode negative (DCEN) are used.

The constant voltage type of direct current power is more popular for submerged arc welding with 1/8 inch (3.2 mm) and smaller diameter electrode wires. The constant current power system is normally used for welding with 5/32 inch(4 mm) and larger-diameter electrode wires. The control circuit for CC power is more complex since it attempts to duplicate the actions of the welder to retain a specific arc length. The wire feed system must sense the voltage across the arc and feed the electrode wire into the arc to maintain this voltage. As conditions change, the wire feed must slow down or speed up to maintain the prefixed voltage across the arc. This adds complexity to the control system. The system cannot react instantaneously. Arc starting is more complicated with the constant current system since it requires the use of a reversing system to strike the arc, retract, and then maintain the preset arc voltage.

For SAW AC welding, the constant current power is always used. When multiple electrode wire systems are used with both AC and DC arcs, the constant current power system is utilized. The constant voltage system, however, can be applied when two wires are fed into the arc supplied by a single power source. Welding current for submerged arc welding can vary from as low as 50 A to as high as 2,000 A. Most submerged arc welding is done in the range of 200 A to 1,200 A.

23.5　Advantages and limitations

Advantages:
- High deposition rates (over 45 kg/h (100 lb/h) have been reported).
- High operating factors in mechanized applications.
- Deep weld penetration.
- Sound welds are readily made (with good process design and control).
- High speed welding of thin sheet steels up to 5 m/min (16 ft/min) is possible.
- Minimal welding fume or arc light is emitted.
- Practically no edge preparation is necessary depending on joint configuration and required penetration.
- The process is suitable for both indoor and outdoor works.

- Welds produced are sound, uniform, ductile, corrosion resistant and have good impact value.
- Single pass welds can be made in thick plates with normal equipment.
- The arc is always covered under a blanket of flux, thus there is no chance of spatter of weld.
- 50% to 90% of the flux is recoverable, recycled and reused.

Limitations:

- Limited to ferrous (steel or stainless steels) and some nickel-based alloys.
- Normally limited to the 1F, 1G, and 2F positions.
- Normally limited to long straight seams or rotated pipes or vessels.
- Requires relatively troublesome flux handling systems.
- Flux and slag residue can present a health and safety concern.
- Requires inter-pass and post weld slag removal.
- Requires backing strips for proper root penetration.
- Limited to high thickness materials.

Words and terms

longitudinal seam　纵向焊缝

heavy industry　重工业

uranium　铀

sound weld　良好的焊缝

arc light　弧光

pipe　管

vessel　船,舰

duplicate　复制,二重的

welding fume　焊接烟尘

post weld slag removal　焊后除渣

Questions

What position can be used when SAW is conducted?

List three specific applications of SAW, and explain why SAW is preferred.

24

Resistance spot welding

Resistance welding is a group of thermo-electric processes in which coalescence is produced by the heat obtained from resistance of the work to electric current in a circuit of which the work is a part and by the application of pressure. The resistant spot welding can be illustrated Fig. 24.1.

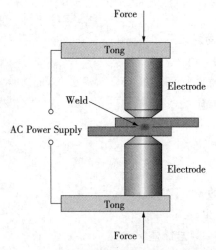

Fig. 24. 1　Diagram of resistance spot welding

Said another way, pressure is applied to the two overlapping sheets being joined. Electrical current is applied causing resistive heating, which results in the melting of metal and the formation of a weld. The weld is called a weld nugget.

There are at least seven important resistance-welding processes. These are:

1) flash welding

2) high frequency

3) percussion welding

4) projection welding

5) resistance seam welding

6) resistance spot welding (most common process), involves the use of water cooled copper electrodes which are clamped with the sheets into place. The electrical current is then applied to the electrodes causing the weld nugget to form.

7) upset welding

The resistance welding processes differ from many of the other more popular welding processes such as MIG, Stick and TIG. Filler metal is rarely used and fluxes are not employed.

Three factors involved are:

1) The amount of current that passes through the work,

2) The pressure that the electrodes transfer to the work,

3) The time the current flows through the work.

Heat is generated by the passage of electrical current through a resistance circuit. The force applied before, during, and after the current flow forces the heated parts together so that coalescence will occur. Pressure is required throughout the entire welding cycle to assure a continuous electrical circuit through the work.

Resistance welds are made very quickly; however, each process has its own time cycle.

Resistance welding operations are automatic. The pressure is applied by mechanical, hydraulic, or pneumatic systems. Motion, when it is involved, is applied mechanically. Current control is completely automatic once the welding operator initiates the weld. Resistance welding equipment utilizes programmers for controlling current, time cycles, pressure, and movement. Welding programs for resistance welding can become quite complex. In view of this, quality welds do not depend on welding operator skill but more on the proper set up and adjustment of the equipment and adherence to weld schedules.

24.1 Applications

This type of welding is used where long production runs and consistent conditions can be maintained. Welding is performed with operators who normally load and unload the welding machine and operate the switch for initiating the weld operation. The automotive industry is the major user of the resistance welding processes, followed by the appliance industry.

Resistance welding is used by many industries manufacturing a variety of products made of thinner gauge metals.

This type of welding is also used in the steel industry for manufacturing pipe, tubing and smaller structural sections. It has the advantage of producing a high volume of work at high speeds and does not require filler materials. Welds are reproducible and high-quality welds are normal.

Metals that are weldable, the thicknesses that can be welded, and joint design are related to specific resistance welding processes.

Most of the common metals can be welded by many of the resistance welding processes (Table 24.1).

Table 24.1 **Weldability of base metals**

Base metal	Weldability
Aluminums	weldable
Magnesium	weldable
Inconel	weldable
Nickel	weldable
Nickel silver	weldable
Monel	weldable
Precious metals	weldable
Low carbon steel	weldable
Low alloy steel	weldable
High & medium carbon	possible, not popular
Steel alloys	possible, not popular
Stainless Steel	weldable

Difficulties may be encountered when welding certain metals in thicker sections. Some metals require heat treatment after welding for satisfactory mechanical properties.

24.2 **How it works**

- Bulk Resistance: Metals have what is called a PTC or Positive Temperature Coefficient. This means that their resistance increases as temperature increases.
- Contact Resistance: When two surfaces come in contact, microscopically the surfaces are rough, where some points come in contact on the surface and some do not. At the points where contact is made and assuming the two pieces of metal are pressed together with some pressure, the oxide layer breaks forming a limited number of metal-to-metal bridges. The weld current is then passed over a large area as it moves through the bulk metal. Since the current is forced through a limited number of bridges this "necking down" increases the current density, generating heat which causes melting. As melting occurs, new bridges are formed. Molten metal also has higher resistance than non-molten metal, forcing the current toward newer bridges. The process proceeds until the entire surface is molten.

When the electrical current is turned off, the electrodes cool first, which then cool the molten metal. When everything solidifies the weld is formed.

Words and terms

hydraulic 液压的,水力学的

appliance industry 器械工业

current density 电流密度

reproducible 可再生的

pneumatic 气胎

satisfactory 满意的,符合要求的

weld nugget 点焊熔核

24.3 Weldability

Weldability is controlled by three factors:

1) Resistivity,

2) Thermal conductivity, and

3) Melting temperature.

Metals with a high resistance to current flow and with a low thermal conductivity and a relatively low melting temperature would be easily weldable. Ferrous metals all fall into this category. Metals that have a lower resistivity but a higher thermal conductivity will be slightly more difficult to weld. This includes the light metals, aluminum and magnesium. The precious metals comprise the third group. These are difficult to weld because of very high thermal conductivity. The fourth group is the refractory metals, which have extremely high melting points and are more difficult to weld.

24.4 Welding heat and current

Welding heat is proportional to the square of the welding current. If the current is doubled, the heat generated is quadrupled. Welding heat is proportional to the total time of current flow, thus, if current is doubled, the time can be reduced considerably. The welding heat generated is directly proportional to the resistance and is related to the material being welded and the pressure applied. The heat losses should be held to a minimum. It is an advantage to shorten welding tire.

Mechanical pressure which forces the parts together helps refine the grain structure of the weld.

Heat is also generated at the contact between the welding electrodes and the work. This amount of heat generated is lower since the resistance between high conductivity electrode material and the normally employed mild steel is less than that between two pieces of mild steel.

In most applications, the electrodes are water cooled to minimize the heat generated between the electrode and the work.

24.5 Types of electrodes

Electrodes vary by shape (called electrode geometry). The right electrode is selected in order to improve electrical-thermal-mechanical performance. As the cross-sectional area increases rapidly with distance from the workpiece, thereby providing a good heat sink.

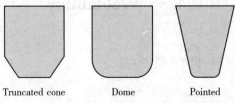

Truncated cone　　　Dome　　　Pointed

Fig. 24. 2　Types of electrodes

The diameter of the electrode contact area is a consideration; if the area is too small, it will produce undersized welds with insufficient strength; if the diameter of the electrode is too large, it will cause inconsistent and unstable weld growth characteristics.

The electrode must be able to:

1) conduct current to the workpiece,

2) mechanically constrain the workpiece, and

3) conduct heat from the workpiece.

The materials used to construct the electrode have to sustain high loads at high temperatures, while maintaining adequate electrical and thermal conductivity.

A range of copper-based or refractory-based electrode materials are used based on the application. Three groups of electrode materials are outlined below. Within each group, the Resistance Welding Manufacturers Association (RWMA) sorts electrode materials into classes.

- Group A contains copper-based alloys. Common examples are:

　　Class 1: (99% copper, 1% cadmium; 60 ksi UTS (forged); conductivity 92% IACS) Specifically recommended, because of its high electrical and thermal conductivity, for spot welding aluminum alloys, magnesium alloys, brass and bronze.

　　Class 2: (99. 2% copper, 0. 8% chromium; 62 ksi UTS (forged), 82% IACS) General purpose electrode material for production spot and seam welding of most materials.

- Group B contains refractory metals and refractory metal composites.

- Group C contains specialty materials such as dispersion-strengthened copper.

In general, higher level of resistance occurs when the power supply overcomes the level of resistance. To produce higher levels of resistance dissimilar parts are used.

- Conductive electrodes such as copper are used to weld resistive materials like nickel or stainless steel.
- Resistive electrodes such as those made from molybdenum are used for welding conductive metals such as gold or copper.

24.6　Types of bonds

- Stainless State Bond (also referred to as a therm-compression bond): Metals such as tungsten and molybdenum, which are dissimilar materials that have a dissimilar grain structure are joined together with high weld energy, a short heating time and a high level of forces. When this occurs there isn't much melting needed to form a bond. Peel strength is poor but tensile and shear strength is high.
- Fusion Bond: With these approaches dissimilar or similar material that have similar grain structures are used. The metal is heated to the melting point of both metals. A "nugget" alloy of both materials is formed with larger grain growth. The bond formed has excellent shear strength, peel and tensile strengths.
- Reflow Braze Bond: In this approach the resistance heating of gold, silver or other low temperature brazing material is joined with either a widely varied thick/thin material or dissimilar materials. The brazing material must "wet" to each part and have a lower melting point than the two workpieces. The process requires a longer heating time (2 ms to 100 ms) at low weld energy. The result is a bond with excellent tensile strength. The shear strength and peel are poor.

24.7　Advantages and disadvantages

Advantages:

The automotive and appliance industries choose resistance welding for manufacturing because of the great advantages this process has to offer. The first advantage is speed. When over 5,000 welds need to be made in a typical car, a process where each weld takes less than a second is of great importance. The process is also adaptable to robotic manipulation so the speed is extremely fast. It is excellent for the sheet metals used in automotive construction, and because no filler metal is needed, the complex wire feed systems in many arc welding processes are avoided.

- Higher speeds, <1 s for automotive spot welding, short process time
- Excellent for sheet metal applications<1/4 inch
- No filler metals or consumables required
- Relatively safe due to low voltage requirements
- Environmentally friendly clean process
- Joint formed is reliable

Disadvantages:

Several disadvantages are associated with this process. Resistance welding equipment is more expensive than arc welding equipment. The process lacks the portability of arc welding. Although individual spot welding guns may have limited movement on the assembly line, the power source is fixed.

Parts to be joined are limited to a thickness of less than 1/4 of an inch due to current requirements. Thicker base materials have a greater ability to dissipate heat away from the weld area. Also, the resistance welding process is limited to overlapping joints, which requires more material than a butt joint.

The process can produce unfavorable power line demands, particularly with single-phase as opposed to 3-phase transformers. Short time, high power demands can cause lights to dim and computers to reset if the electrical system in a factory is not properly prepared for the introduction of resistance welding equipment.

The lack of a simple, in-process nondestructive testing technique for resistance spot and seam welding is also a limitation. Because resistance welds are produced between overlapping sheets, there can be no visual examination if the finished weld. Also, the time required for ultrasonic inspection of individual spot welds would be unacceptable in a high production environment such as the automotive industry.

Spot welds have low tensile and fatigue strength; the notch around the periphery of the nugget between the sheets acts as a stress concentrator.

Electrode wear acts to increase the diameter of the electrode face. During production, current values must slowly rise to compensate for the decreased current density, else nugget size drops.

- Power requirements
- Nondestructive testing
- Low fatigue and tensile strength
- Is not portable
- High levels of electrode wear
- If uses a lap joint, it requires additional metal

Words and terms

resistivity　电阻率,抵抗力

category　种类分类

electrode geometry　电极几何形状

sustain　维持,承担

robotic manipulation　自动化操作

periphery　外围,边缘,圆周

thermal conductivity　热导率

mild steel　低碳钢

heat sink　散热器

refractory　难熔的,耐火物质

nondestructive　无损的,非破坏性的

Questions

Which of the following alloys can be easily welded by resistance spot welding: steel, aluminum or copper? Why?

What type of bond will the brazing between dissimilar metals form? Why?

25

Resistance seam welding

Resistance seam welding is a process that produces a weld at the faying surfaces of two similar metals. The seam may be a butt joint or an overlap joint and is usually an automated process. It differs from butt welding in that butt welding typically welds the entire joint at once and seam welding forms the weld progressively, starting at one end. Like spot welding, seam welding relies on two electrodes, usually made from copper, to apply pressure and current. The electrodes are often disc shaped and rotate as the material passes between them. This allows the electrodes to stay in constant contact with the material to make long continuous welds. The electrodes may also move or assist the movement of the material. The resistant seam welding process can be illustrated in Fig. 25.1.

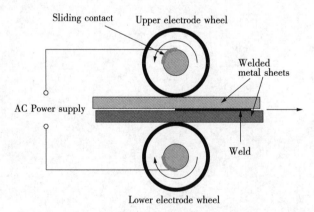

Fig. 25.1　Diagram of resistance seam welding

A transformer supplies energy to the weld joint in the form of low voltage, high current AC power. The joint of the work piece has high electrical resistance relative to the rest of the circuit and is heated to its melting point by the current. The semi-molten surfaces are pressed together by the welding pressure that creates a fusion bond, resulting in a uniformly welded structure. Most seam welders use water cooling through the electrode, transformer and controller assemblies due to the heat generated.

Seam welding produces an extremely durable weld because the joint is forged due to the heat and pressure applied. A properly welded joint formed by resistance welding can easily be stronger than the material from which it is formed.

A common use of seam welding is during the manufacture of round or rectangular steel tubing. Seam welding has been used to manufacture steel beverage cans but is no longer used for this as modern beverage cans are seamless aluminum.

There are two modes for seam welding: Intermittent and continuous. In intermittent seam welding, the wheels advance to the desired position and stop to make each weld. This process continues until the desired length of the weld is reached. In continuous seam welding, the wheels continue to roll as each weld is made.

Words and terms

bulk resistance　体电阻

overlap joint　搭接接头

quadruple　四重的

fatigue strength　疲劳强度

cathode　阴极

intermittent　间歇的

faying surface　结合面

semi-molten　半熔化

contact resistance　接触电阻

anode　阳极

beverage can　饮料罐头

Questions

Why does the seam welding produce an extremely durable weld?

What kind of microstructural evolution will occur during seam welding of aluminum?

26

Electroslag welding

➢➢➢➢➢➢➢➢➢➢➢➢➢➢➢➢➢➢➢➢

Electroslag welding (ESW) is a highly productive, single pass welding process for thick (greater than 25 mm up to about 300 mm) materials in a vertical or close to vertical position. ESW is similar to electro gas welding, but the main difference is the arc starts in a different location. As shown in Fig. 26.1, an electric arc is initially struck by wire that is fed into the desired weld location and then flux is added. Additional flux is added until the molten slag, reaching the tip of the electrode, extinguishes the arc. The wire is then continually fed through a consumable guide tube (can oscillate if desired) into the surfaces of the metal workpieces and the filler metal are then melted using the electrical resistance of the molten slag to cause coalescence. The wire and tube then move up along the workpiece while a copper retaining shoe that was put into place before starting (can be water-cooled if desired) is used to keep the weld between the plates that are being welded. Electroslag welding is used mainly to join low carbon steel plates and/or sections that are very thick. It can also be used on structural steel if certain precautions are observed. This process uses a direct current (DC) voltage usually ranging from

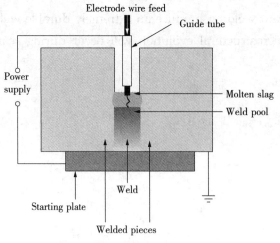

Fig. 26.1　Diagram of electroslag welding

about 600 A and 40V to 50 V, higher currents are needed for thicker materials. Because the arc is extinguished, this is not an arc process.

26.1　History

The process was patented by Robert K Hopkins in the United States in February 1940 (patent 2191481) and developed and refined at the Paton Institute, Kiev, USSR during the 1940s. The Paton method was released to the west at the Bruxelles Trade Fair of 1950. The first widespread use in the U.S. was in 1959, by General Motors Electromotive Division, Chicago, for the fabrication of traction motor frames. In 1968 Hobart Brothers of Troy, Ohio, released a range of machines for use in the shipbuilding, bridge construction and large structural fabrication industries. Between the late 1960s and late 1980s, it is estimated that in California alone over a million stiffeners were welded with the Electroslag welding process. Two of the tallest buildings in California were welded, using the Electroslag welding process—The Bank of America building in San Francisco, and the twin tower Security Pacific buildings in Los Angeles. The Northridge earthquake and the Loma Prieta earthquakes provided a "real world" test to compare all of the welding processes. The Structural Steel welding industry is well aware that, over one billion dollars in crack repairs were needed, after the Northridge earthquake, to repair weld cracks propagated in welds made with the gasless flux cored wire process. Not one failure or one crack propagation was initiated in any of the hundreds-of-thousands of welds made on continuity plates welded with the Electroslag welding process.

However the Federal Highway Administration (FHWA) monitored the new process and found that Electroslag welding, because of the very large amounts of confined heat used, produced a coarse-grained and brittle weld and in 1977 banned the use of the process for many applications. The FHWA commissioned research from universities and industry and Narrow Gap Improved Electroslag Welding (NGI-ESW) was developed as a replacement. The FHWA moratorium was rescinded in 2000.

26.2　Benefits

Benefits of the process include its high metal deposition rates—it can lay metal at a rate between 15 kg/h and 20 kg/h (35 lb/h and 45 lb/h) per electrode—and its ability to weld thick materials. Many welding processes require more than one pass for welding thick workpieces, but often a single pass is sufficient for electroslag welding. The process is also very efficient, since joint preparation and materials handling are minimized while filler metal utilization is high. The process is also safe and clean, with no arc flash and low weld splatter or distortion. Electroslag welding easily lends itself to mechanization, thus reducing the requirement for skilled manual welders.

One electrode is commonly used to make welds on materials with a thickness of 25 mm to 75 mm(1inch to 3 inch), and thicker pieces generally require more electrodes. The maximum workpiece thickness that has ever been successfully welded was a 0.91 m (36 inch) piece that required the simultaneous use of six electrodes to complete.

27

Exothermic welding

27. 1 Welding principle

Exothermic welding, also known as exothermic bonding, thermite welding (TW), and thermit welding, is a welding that employs molten metal to permanently join the conductors. The process employs an exothermic reaction of a thermite composition to heat the metal, and requires no external source of heat or current, as shown in Fig. 27. 1. The chemical reaction that produces the heat is an aluminothermic reaction between aluminium powder and a metal oxide.

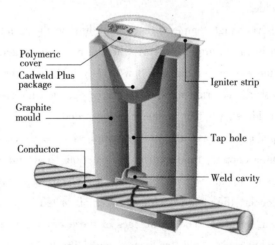

Fig. 27. 1 Diagram of exothermic welding

In exothermic welding, aluminium dust reduces the oxide of another metal, most commonly iron oxide, because aluminium is highly reactive. Iron(III) oxide is commonly used:

$$Fe_2O_3 + 2Al \longrightarrow 2Fe + Al_2O_3$$

The products are aluminium oxide, free elemental iron, and a large amount of heat. The reactants are commonly powdered and mixed with a binder to keep the material solid and prevent separation.

Commonly the reacting composition is five parts iron oxide red (rust) powder and three parts aluminium powder by weight, ignited at high temperatures. A strongly exothermic (heat-generating) reaction occurs that via reduction and oxidation produces a white hot mass of molten iron and a slag of refractory aluminium oxide. The molten iron is the actual welding material; the aluminium oxide is much less dense than the liquid iron and so floats to the top of the reaction, so the set-up for welding must take into account that the actual molten metal is at the bottom of the crucible and covered by floating slag.

Other metal oxides can be used, such as chromium oxide, to generate the given metal in its elemental form. Copper thermite, using copper oxide, is used for creating electric joints:

$$3Cu_2O+2Al \longrightarrow 6Cu+Al_2O_3$$

Thermite welding is widely used to weld railway rails. One of the first railroads to evaluate the use of thermite welding was the Delaware and Hudson Railroad in the United States in 1935. The weld quality of chemically pure thermite is low due to the low heat penetration into the joining metals and the very low carbon and alloy content in the nearly pure molten iron. To obtain sound railroad welds, the ends of the rails being thermite welded are preheated with a torch to an orange heat, to ensure the molten steel is not chilled during the pour. Because the thermite reaction yields relatively pure iron, not the much stronger steel, some small pellets or rods of high-carbon alloying metal are included in the thermite mix; these alloying materials melt from the heat of the thermite reaction and mix into the weld metal. The alloying beads composition will vary, according to the rail alloy being welded.

The reaction reaches very high temperatures, depending on the metal oxide used. The reactants are usually supplied in the form of powders, with the reaction triggered using a spark from a flint lighter. The activation energy for this reaction is very high however, and initiation requires either the use of a "booster" material such as powdered magnesium metal or a very hot flame source. The aluminium oxide slag that it produces is discarded.

When welding copper conductors, the process employs a semi-permanent graphite crucible mould, in which the molten copper, produced by the reaction, flows through the mould and over and around the conductors to be welded, forming an electrically conductive weld between them. When the copper cools, the mould is either broken off or left in place. Alternatively, hand-held graphite crucibles can be used. The advantages of these crucibles include portability, lower cost (because they can be reused), and flexibility, especially in field applications.

27.2 Applications

Exothermic welding is usually used for welding copper conductors but is suitable for welding a wide range of metals, including stainless steel, cast iron, common steel, brass, bronze, and Monel. It is especially useful for joining dissimilar metals. The process is marketed under a variety of names such as APLIWELD (in tablet form), American Rail Weld, Harger ULTRASHOT, ERICO CADWELD, Quikweld, Tectoweld, Ultraweld, Techweld, TerraWeld, Thermoweld, Ardo Weld, AmiableWeld, AIWeld, FurseWeld and Kumwell.

Because of the good electrical conductivity and high stability in the face of short-circuit pulses, exothermic welds are one of the options specified by §250.7 of the United States National Electrical Code for grounding conductors and bonding jumpers. It is the preferred method of bonding, and indeed it is the only acceptable means of bonding copper to galvanized cable. The NEC does not require such exothermically welded connections to be listed or labelled, but some engineering specifications require that completed exothermic welds be examined using X-ray equipment.

27.3　Rail welding

Typically, the ends of the rails are cleaned, aligned flat and true, and spaced apart 25 mm (1 inch). This gap between rail ends for welding is to ensure consistent results in the pouring of the molten steel into the weld mold. In the event of a welding failure, the rail ends can be cropped to a 75 mm (3 inch) gap, removing the melted and damaged rail ends, and a new weld attempted with a special mould and larger thermite charge. A two or three piece hardened sand mould is clamped around the rail ends, and a torch of suitable heat capacity is used to preheat the ends of the rail and the interior of the mould. The proper amount of thermite with alloying metal is placed in a refractory crucible, and when the rails have reached a sufficient temperature, the thermite is ignited and allowed to react to completion (allowing time for any alloying metal to fully melt and mix, yielding the desired molten steel or alloy). The reaction crucible is then tapped at the bottom. Modern crucibles have a self-tapping thimble in the pouring nozzle. The molten steel flows into the mould, fusing with the rail ends and forming the weld. The slag, being lighter than the steel flows last from the crucible and overflows the mould into a steel catch basin, to be disposed of after cooling. The entire setup is allowed to cool. The mould is removed and the weld is cleaned by hot chiselling and grinding to produce a smooth joint. Typical time from start of the work until a train can run over the rail is approximately 45 min to more than an hour, depending on the rail size and ambient temperature. In any case, the rail steel must be cooled to less than 370 ℃ (700 ℉) before it can sustain the weight of rail locomotives.

When a thermite process is used for track circuits - the bonding of wires to the rails with a copper alloy, a graphite mould is used. The graphite mould is reusable many times, because the copper alloy is not as hot as the steel alloys used in rail welding. In signal bonding, the volume of molten copper is quite small, approximately 2 cm^3 and the mould is lightly clamped to the side of the rail, also holding a signal wire in place. In rail welding, the weld charge can weigh up to 13 kg (29 lb). The hardened sand mould is heavy and bulky, must be securely clamped in a very specific position and then subjected to intense heat for several minutes before firing the charge. When rail is welded into long strings, the longitudinal expansion and contraction of steel must be taken into account. British practice sometimes uses a sliding joint of some sort at the end of long runs of continuously welded rail, to allow some movement, although by using a heavy concrete sleeper and an extra amount of ballast at the sleeper ends, the track, which will be prestressed according to the ambient temperature at the time of its installation, will develop compressive stress in hot ambient temperature, or tensile stress in cold ambient temperature, its strong attachment to the heavy sleepers preventing buckling or other deformation. Current practice is to use welded rails throughout on high speed lines, and expansion joints are kept to a minimum, often only to protect junctions and crossings from excessive stress. American practice appears to be very similar, a straightforward physical restraint of the rail. The rail is prestressed, or considered "stress neutral" at some particular ambient temperature. This "neutral" temperature will vary according to local climate conditions, taking into account lowest winter and warmest summer temperatures. The rail is physically secured to the ties or sleepers with rail anchors, or anti-creepers. If the track ballast is good and clean and the ties are in good condition, and the track geometry is good, then the welded rail will withstand ambient temperature swings normal to the region.

Words and terms

thermit welding　铝热焊

electroslag welding　电渣焊

minimize　最小化

aluminothermic reaction　铝热反映

flexibility　灵活性,可行性

stability　稳定性,坚定

locomotive　火车头,火车头的,运动的,移动的

track circuit　轨道电路

ballast　压载,包袱

ambient temperature　室温

exothermic welding　铝热焊,放热焊

extinguish　熄灭,压制

crucible　坩埚

chill　冷却,冷冻,寒冷的

pellet　小球

chisel 凿子,雕,刻
prestress 预应力
neutral 中立的,中性的
swing 摇摆,摆动

Questions

Why is the weld quality of chemically pure thermite low? How to solve this problem?

In what situation the thermit welding is preferred? What kind of other welding method is likely to replace thermit welding in some cases? Why?

28

Laser beam welding

28.1 Overview

Laser beam welding is a technique in manufacturing whereby two or more pieces of material (usually metal) are joined by together through use of a laser beam, as illustrated in Fig. 28. 1. Laser stands for Light Amplification by Stimulated Emission of Radiation. It is a non-contact process that requires access to the weld zone from one side of the parts being welded.

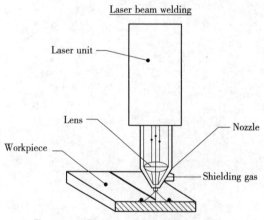

Fig. 28. 1   Diagram of laser beam welding

The weld is formed as the intense laser light rapidly heats the material - typically calculated in Milli-seconds.

The laser beam is a coherent (single phase) light of a single wavelength (monochromatic). The laser beam has low beam divergence and high energy content and thus will create heat when it strikes a surface.

The primary types of lasers used in welding and cutting are:

- Gas lasers: use a mixture of gases such as helium and nitrogen. There are also CO_2 or carbon dioxide lasers. These lasers use a low-current, high-voltage power source to excite the gas mixture using a lasing medium. Operate in a pulsed or continuous mode. Carbon dioxide lasers use a mixture of high purity carbon dioxide with helium and nitrogen as the lasing medium. CO_2 lasers are also used in dual beam laser welding where the beam is split into two equal power beams.

- Solid state lasers: (Nd: YAG type and ruby lasers) Operate at 1micrometer wavelengths. They can be pulsed or operate continuously. Pulsed operation produced joints similar to spot welds but with complete penetration. The pulse energy is 1 J to 100 J. Pulse time is 1ms to 10 ms.

- Diode lasers: Lasers are used for materials that are difficult to weld using other methods, for hard to access areas and for extremely small components. Inert gas shielding is needed for more reactive materials.

28.2　History

Einstein first postulated the quantum-mechanical fundamentals of lasers at the beginning of the 20th century.

The first laser called a ruby laser was first implemented in 1960.

The first high performance lasers were developed in the 1970s with the development of CO_2 lasers. Since this time the applications for laser beam sources have evolved.

Laser soldering becomes a popular way to join leads in electronic components through holes in printed circuit boards in 1980.

Laser powder fusion process developed in 1987.

In 2002 from Linde Gas in Germany, a Diode laser using process gases and "active-gas components" is investigated to enhance the "key-holing" effects for laser welding. The process gas, Argon-CO_2, increases the welding speed and in the case of a diode laser, will support the transition of heat conductivity welding to a deep welding, i. e., 'key-holing'. Adding active gas changes the direction of the metal flow within a weld pool and produces narrower, high-quality weld.

CO_2 Lasers are used to weld polymers. The Edison Welding Institute is using through-transmission lasers in the range of 230 nm to 980 nm to readily form welded joints. Using silicon carbides embedded in the surfaces of the polymer, the laser is capable of melting the material leaving a near invisible joint line.

Words and terms

stimulate　刺激,鼓励,激励

radiation　辐射,发光

emission　发射,散射

divergence　分歧

polymer　聚合物

postulate　假定,要求,基本条件

diode laser　二极管激光器

invisible　无形的,看不见的

solid state laser　固体激光器

Questions

List the main components of laser beam equipment.

When is shielding gas necessary in laser beam welding?

28.3　Laser welding vs. arc welding

Laser beam welding energy transfer is different than arc welding processes. In laser welding the absorption of energy by a material is affected by many factors such as the type of laser, the incident power density and the base metal's surface condition.

Laser output is not electrical in nature and does not require a flow of electrical current. This eliminates any effect of magnetism, and does not limit the process to electrically conductive materials.

Lasers can interact with any material. It doesn't require a vacuum and it does not produce x-rays.

28.4　Welding principle

- Pump source provides energy to the medium, exciting the laser such that electrons held within the atoms are elevated temporarily to higher energy states.
- The electrons held in this excited state cannot remain there indefinitely and drop down to a lower energy level.
- The electron loses the excess energy gained from the pump energy by emitting a photon. This is called spontaneous emission and the photons produced by this method are the seed for laser generation.
- Photons emitted by spontaneous emission eventually strike other electrons in the higher energy states. The incoming photon "knocks" the electron from the excited state to a lower

energy level creating another photon. These photons are coherent meaning they are in phase, of the same wavelength, and traveling the same direction. A process called stimulated emission.

- Photons are emitted in all directions, however some travel along the laser medium to strike the resonator mirrors to be reflected back through the medium. The resonator mirrors define the preferential amplification direction for stimulated emission. In order for the amplification to occur there must be a greater percentage of atoms in the excited state than the lower energy levels. This population inversion of more atoms in the excited state leads to the conditions required for laser generation.

- The focus spot of the laser is targeted on the workpiece surface which will be welded. At the surface the concentration of light energy converts into thermal energy (heat). The heat causes the surface of the material to melt, which progresses through the surface by a process called surface conductivity. The beam energy level is maintained below the vaporization temperature of the workpiece material. The ideal thickness of the materials to be welded is 20 mm. The energy is a laser is concentrated, an advantage when working with materials that have high thermal conductivity.

28.5　Welding characteristics

The laser can be compared to solar light beam for welding. It can be used in air. The laser beam can be focused and directed by special optical lenses and mirrors. It can operate at considerable distance from the workpiece.

When using the laser beam for welding, the electromagnetic radiation impinges on the surface of the base metal with such a concentration of energy that the temperature of the surface is melted vapor and melts the metal below. One of the original questions concerning the use of the laser was the possibility of reflectivity of the metal so that the beam would be reflected rather than heat the base metal. It was found, however, that once the metal is raised to its melting temperature, the surface conditions have little or no effect.

The distance from the optical cavity to the base metal has little effect on the laser. The laser beam is coherent and it diverges very little. It can be focused to the proper spot size at the work with the same amount of energy available, whether it is close or far away.

With laser welding, the molten metal takes on a radial configuration similar to convectional arc welding. However, when the power density rises above a certain threshold level, keyholing occurs, as with plasma arc welding. Keyholing provides for extremely deep penetration. This provides for a high depth-to-width ratio. Keyholing also minimizes the problem of beam reflection from the shiny molten metal surface since the keyhole behaves like a black body and absorbs the majority of the energy. In some applications, inert gas is used to shield the molten metal from the atmosphere. The metal vapor that occurs may cause a breakdown of the shielding gas and creates plasma in the region of high-beam intensity just above the metal surface. The plasma absorbs energy from the laser beam and can actually block the beam and reduce melting. Use of an inert gas jet directed along the metal surface eliminates the plasma buildup and shields the surface from the atmosphere.

The welding characteristics of the laser and of the electron beam are similar. The concentration of energy by both beams is similar with the laser having a power density in the order of 106 w/cm^2. The power density of the electron beam is only slightly greater. This is compared to a current density of only 104 w/cm^2 for arc welding.

Laser beam welding has a tremendous temperature differential between the molten metal and the base metal immediately adjacent to the weld. Heating and cooling rates are much higher in laser beam welding than in arc welding, and the heat-affected zones are much smaller. Rapid cooling rates can create problems such as cracking in high carbon steels.

Experimental work with the laser beam welding process indicates that the normal factors control the weld. Maximum penetration occurs when the beam is focused slightly below the surface. Penetration is less when the beam is focused on the surface or deep within the surface. As power is increased the depth of penetration is increased.

28.6 Types of welds

Performed at lower energy levels forming a wide and shallow weld nugget. There are two modes:

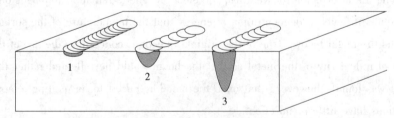

Notes: 1-Conduction mode welding; 2-Conduction/penetration mode; 3-Penetration or keyhole mode
Fig. 28.2 Types of welds

1) Direct heating: heat flow is governed by classical thermal conduction from a surface heat source. The weld is made by melting portions of the base material. Can be made using pulsed ruby and CO_2 lasers using a wide range of alloys and metals. Can also use Nd: YAD and diode lasers.

2) Energy transmission: energy is absorbed through novel inter-facial absorption methods. An absorbing ink is placed at the interface of a lap joint. The ink absorbs the laser beam energy,

which is conducted into a limited thickness of surrounding material to form a molten inter-facial film that solidifies as the welded joint. Butt welds can be made by directing the energy towards the joint line at an angle through material at one side of the joint, or from one end if the material is highly transmissive.

Conduction/penetration welding occurs at medium energy density and results in more penetration.

The keyhole mode welding creates deep narrow welds. In this type of welding the laser light forms a filament of vaporized material known as a "keyhole" that extends into the material and provides conduit for the laser light to be efficiently delivered into the material.

The direct delivery of energy into the material does not rely on conduction to achieve penetration, so it minimizes the heat into the material and reduces the heat affected zone.

28.7　Applications

The laser beam has been used to weld: carbon steels, high strength low alloy steels, aluminum, stainless steel, titanium.

Laser welds made in these materials are similar in quality to welds made in the same materials by electron beam process. Experimental work using filler metal is being used to weld metals that tend to show porosity when welded with either EB or LB welding. Materials 1/2 inch. (12.7 mm) thick are being welded at a speed of 10.0 inch/min (254.0 mm/min).

28.8　Advantages and disadvantages

Advantages:
- Works with high alloy metals without difficulty,
- Can be used in open air,
- Can be transmitted over long distances with a minimal loss of power,
- Narrow heat affected zone,
- Low total thermal input,
- Welds dissimilar metals,
- No filler metals necessary,
- No secondary finishing necessary,
- Extremely accurate,
- Produces deep and narrow welds,
- Low distortion in welds,

- High quality welds,
- Can weld small, thin components,
- No contact with materials.

Disadvantages:

- Rapid cooling rate may cause cracking in some metals,
- High capital cost for equipment,
- Optical surfaces of the laser are easily damaged,
- High maintenance costs.

Words and terms

absorption 吸收,全神贯注

magnetism 磁性,磁力

photon 光子,光量子

conduction mode 导电方式

eliminate 消除,排除

pump source 泵浦源;泵源

resonator mirror 共振腔反射镜

dissimilar metal 异种金属

Questions

Under what circumstances laser welding is more preferred than TIG welding? Why? Please analyze the common welding defects and causes in the laser welding process.

29

Electron beam welding

29.1 Overview

Electron beam welding (EBW) is a fusion welding process in which a beam of high-velocity electrons is applied to two materials to be joined. The workpieces melt and flow together as the kinetic energy of the electrons is transformed into heat upon impact. EBW is often performed under vacuum conditions to prevent dissipation of the electron beam, as shown in Fig. 29.1.

Electron beam welding was developed by the German physicist Karl-Heinz Steigerwald, who was at the time working on various electron beam applications. Steigerwald conceived and developed the first practical electron beam welding machine, which began operation in 1958. American inventor James T. Russell has also been credited with designing and building the first electron-beam welder.

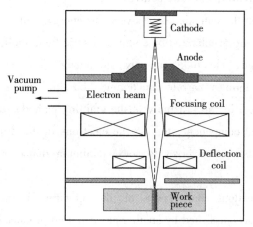

Fig. 29.1   Diagram of electron beam welding

29.2　Beam formation

Cathode — the source of free electrons

Conduction electrons (those not bound to the nucleus of atoms) move in a crystal lattice of metals with velocities distributed according to Gauss's law and depending on temperature. They cannot leave the metal unless their kinetic energy (in eV) is higher than the potential barrier at the metal surface. The number of electrons fulfilling this condition increases exponentially with increasing temperature of the metal, following Richardson's rule.

As a source of electrons for electron beam welders, the material must fulfill certain requirements: to achieve high power density in the beam, the emission current density $[A/mm^2]$, hence the working temperature, should be as high as possible, to keep evaporation in vacuum low, the material must have a low enough vapor pressure at the working temperature.

The emitter must be mechanically stable, not chemically sensitive to gases present in the vacuum atmosphere (like oxygen and water vapor), easily available, etc.

These and other conditions limit the choice of material for the emitter to metals with high melting points, practically to only two: tantalum and tungsten. With tungsten cathodes, emission current densities about $100 \ mA/mm^2$ can be achieved, but only a small portion of the emitted electrons takes part in beam formation, depending on the electric field produced by the anode and control electrode voltages. The type of cathode most frequently used in electron beam welders is made of a tungsten strip, about 0.05 mm thick, shaped as shown in Fig. 29.1. The appropriate width of the strip depends on the highest required value of emission current. For the lower range of beam power, up to about 2 kW, the width $w = 0.5$ mm is appropriate.

Acceleration of electrons, current control

Electrons emitted from the cathode possess very low energy, only a few eV. To give them the required high speed, they are accelerated by a strong electric field applied between the emitter and another, positively charged, electrode, namely the anode. The accelerating field must also navigate the electrons to form a narrow converging "bundle" around the axis. This can be achieved by an electric field in the proximity of the emitting cathode surface which has, a radial addition as well as an axial component, forcing the electrons in the direction of the axis. Due to this effect, the electron beam converges to some minimum diameter in a plane close to the anode.

For practical applications the power of the electron beam must, of course, be controllable. This can be accomplished by another electric field produced by another cathode negatively charged with respect to the first.

At least this part of electron gun must be evacuated to "high" vacuum, to prevent "burning" the cathode and the emergence of electrical discharges.

Focusing

After leaving the anode, the divergent electron beam does not have a power density sufficient for welding metals and has to be focused. This can be accomplished by a magnetic field produced by electric current in a cylindrical coil.

Magnetic lens

The focusing effect of a rotationally symmetrical magnetic field on the trajectory of electrons is the result of the complicated influence of a magnetic field on a moving electron. This effect is a force proportional to the induction B of the field and electron velocity v. The vector product of the radial component of induction Br and axial component of velocity v_a is a force perpendicular to those vectors, causing the electron to move around the axis. Additional effect of this motion in the same magnetic field is another force F oriented radially to the axis, which is responsible for the focusing effect of the magnetic lens. The resulting trajectory of electrons in the magnetic lens is a curve similar to a helix. In this context it should be mentioned that variations of focal length (exciting current) cause a slight rotation of the beam cross-section.

Beam deflection system

As mentioned above, the beam spot should be very precisely positioned with respect to the joint to be welded. This is commonly accomplished mechanically by moving the workpiece with respect to the electron gun, but sometimes it is preferable to deflect the beam instead. Most often a system of four coils positioned symmetrically around the gun axis behind the focusing lens, producing a magnetic field perpendicular to the gun axis, is used for this purpose.

There are more practical reasons why the most appropriate deflection system is used in TV CRT or PC monitors. This applies to both the deflecting coils as well as to the necessary electronics. Such a system enables not only "static" deflection of the beam for the positioning purposes mentioned above, but also precise and fast dynamic control of the beam spot position by a computer. This makes it possible, e. g. : to weld joints of complicated geometry, to create image-enlarged pictures of objects in the working chamber on TV or PC monitors.

Both possibilities find many useful applications in electron beam welding practice.

Words and terms

converge　集中,聚集

cylindrical coil　圆柱形线圈

trajectory　轨道,轨线

helix　螺旋

symmetrical　均匀的,匀称的

divergent　发散的,相异的

magnetic field　磁场

deflection system　射偏转系统
focal length　焦距
monitor　监视器

Questions

Why is vacuum necessary in EBW?

Please try to analyze the difference between laser beam welding and electron beam welding.

Whether the electron beam welding is harmful to the human body? Why? And if the answer is yes, how to avoid it?

29.3　Penetration of electron beam

To explain the capability of the electron beam to produce deep and narrow welds, the process of "penetration" must be explained. First of all, the process for a "single" electron can be considered.

Penetration of electrons

When electrons from the beam impact the surface of a solid, some of them may be reflected (as "backscattered" electrons), while others penetrate the surface, where they collide with the particles of the solid. In non-elastic collisions they lose their kinetic energy. It has been proved, both theoretically and experimentally, that they can "travel" only a very small distance below the surface before they transfer all their kinetic energy into heat. This distance is proportional to their initial energy and inversely proportional to the density of the solid. Under conditions usual in welding practice the "travel distance" is on the order of hundredths of a millimeter. Just this fact enables, under certain conditions, fast beam penetration.

Penetration of the electron beam

The heat contribution of single electrons is very small, but the electrons can be accelerated by very high voltages, and by increasing their number (the beam current) the power of the beam can be increased to any desired value. By focusing the beam onto a small diameter on the surface of a solid object, values of planar power density as high as 10^4 W/mm^2 up to 10^7 W/mm^2 can be reached. Because electrons transfer their energy into heat in a very thin layer of the solid, as explained above, the power density in this volume can be extremely high. The volume density of power in the small volume in which the kinetic energy of the electrons is transformed into heat can reach values of the order 10^5 W/mm^3 to 10^7 W/mm^3. Consequently, the temperature in this volume increases extremely rapidly, by 108 K/s to 109 K/s.

The effect of the electron beams under such circumstances depends on several conditions, first of all on the physical properties of the material. Any material can be melted, or even evaporated, in a very short time. Depending on conditions, the intensity of evaporation may vary, from negligible to essential. At lower values of surface power density (in the range of about 10^3 W/mm^2) the loss of material by evaporation is negligible for most metals, which is favorable for welding. At higher power density, the material affected by the beam can totally evaporate in a very short time; this no longer electron beam welding; it is electron beam machining.

29.4 Weldability

For welding thin-walled parts, appropriate welding aids are generally needed. Their construction must provide perfect contact of the parts and prevent their movement during welding. Usually they have to be designed individually for a given workpiece.

Not all materials can be welded by an electron beam in a vacuum. This technology cannot be applied to materials with high vapor pressure at the melting temperature, like zinc, cadmium, magnesium and practically all nonmetals.

Another limitation to weldability may be the change of material properties induced by the welding process, such as a high speed of cooling. As detailed discussion of this matter exceeds the scope of this article, the reader is recommended to seek more information in the appropriate literature.

29.5 Joining dissimilar materials

It is often not possible to join two metal components by welding, i. e. to melt part of both in the vicinity of the joint, if the two materials have very different properties from their alloy, due to the creation of brittle, inter-metallic compounds. This situation cannot be changed, even by electron beam heating in vacuum, but this nevertheless makes it possible to realize joints meeting high demands for mechanical compactness and that are perfectly vacuum-tight. The principal approach is not to melt both parts, but only the one with the lower melting point, while the other remains solid. The advantage of electron beam welding is its ability to localize heating to a precise point and to control exactly the energy needed for the process. A high-vacuum atmosphere substantially contributes to a positive result. A general rule for construction of joints to be made this way is that the part with the lower melting point should be directly accessible for the beam.

29.6 Possible problems and limitations

The material melted by the beam shrinks during cooling after solidification, which may have unwanted consequences like cracking, deformation and changes of shape, depending on conditions.

The butt weld of two plates results in bending of the weldment because more material has been melted at the head than at the root of the weld. This effect is of course not as substantial as in arc welding.

Another potential danger is the emergence of cracks in the weld. If both parts are rigid, the shrinkage of the weld produces high stress in the weld which may lead to cracks if the material is brittle (even if only after remelting by welding). The consequences of weld contraction should always be considered when constructing the parts to be welded.

Words and terms

Backscattered   背散射的

localize   使……局部化/停留在某处

inter-metallic compound   金属间化合物

inversely proportional to   反比于……

essential   基本的,必要的

negligible   微不足道的,可以忽略的

accelerate   使……加速

substantial   实质的,大量的

collide   碰撞

weldability   焊接性

literature   文献,著作

Questions

Please conclude the characteristics of electron beam welding.

What kinds of alloys are suitable for electron beam welding?

30

Friction welding

30. 1 Overview

Friction welding was first developed in the Soviet Union, with first experiments taking place in 1956. The American companies Caterpillar, Rockwell International, and American Manufacturing Foundry all developed machines for this process. Patents were also issued throughout Europe and the former Soviet Union. The most extensive historical records are kept with the American Welding Society.

Friction welding is a solid state welding process which produces coalescence of materials by the heat obtained from mechanically-induced sliding motion between rubbing surfaces. The work parts are held together under pressure. This process usually involves the rotating of one part against another to generate frictional heat at the junction.

When a suitable high temperature has been reached, rotational notion ceases. Additional pressure is applied and coalescence occurs.

There are two process variations:

1) In the original process, one part is held stationary and the other part is rotated by a motor which maintains an essentially constant rotational speed. The two parts are brought in contact under pressure for a specified period of time with a specific pressure. Rotating power is disengaged from the rotating piece and the pressure is increased. When the rotating piece stops, the weld is completed. This process can be accurately controlled when speed, pressure, and time are closely regulated.

2) The other variation is inertia welding. A flywheel is revolved by a motor until a preset speed is reached. It, in turn, rotates one of the pieces to be welded. The motor is disengaged from the flywheel and the other part to be welded is brought in contact under pressure with the ro-

tating piece. During the predetermined time during which the rotational speed of the part is reduced, the flywheel is brought to an immediate stop. Additional pressure is provided to complete the weld.

Both methods utilize frictional heat and produce welds of similar quality. Slightly better control is claimed with the original process. The two methods are similar, offer the same welding advantages.

30.2 Key factors

1) The rotational speed which is related to the material to be welded and the diameter of the weld at the interface.

2) The pressure between the two parts to be welded. Pressure changes during the weld sequence. At the start, pressure is very low, but is increased to create the frictional heat. When the rotation is stopped, pressure is rapidly increased so forging takes place immediately before or after rotation is stopped.

3) The welding time is related to the shape and the type of metal and the surface area. It is normally a matter of a few seconds. The actual operation of the machine is automatic. It is controlled by a sequence controller, which can be set according to the weld schedule established for the parts to be joined.

Normally, one of the parts to be welded is round in cross section. This is not an absolute necessity. Visual inspection of weld quality can be based on the flash, which occurs around the outside perimeter of the weld. This flash will usually extend beyond the outside diameter of the parts and will curl around back toward the part but will have the joint extending beyond the outside diameter of the part.

If the flash sticks out relatively straight from the joint, it indicates that the welding time was too short, the pressure was too low, or the speed too high. These joints may crack. If the flash curls too far back on the outside diameter, it indicates that the time was too long and the pressure was too high. Between these extremes is the correct flash shape. The flash is normally removed after welding.

30.3 Types of friction welding

30.3.1 Spin welding

Spin welding systems consist of two chucks for holding the materials to be welded, one of which is fixed and the other rotating. This process can be illustrated in Fig. 30.1. Before welding one of the work pieces is attached to the rotating chuck along with a flywheel of a given weight. The piece is then spun up to a high rate of rotation to store the required energy in the flywheel. Once spinning at the proper speed, the motor is removed and the pieces forced together under pressure. The force is kept on the pieces after the spinning stops to allow the weld to "set".

In Inertia Friction Welding the drive motor is disengaged, and the work pieces are forced together by a friction welding force. The kinetic energy stored in the rotating flywheel is dissipated as heat at the weld interface as the flywheel speed decreases.

In Direct Drive Friction welding the drive motor and chuck are connected. The drive motor is continually driving the chuck during the heating stages. Usually a clutch is used to disconnect the drive motor from the chuck and a brake is then used to stop the chuck.

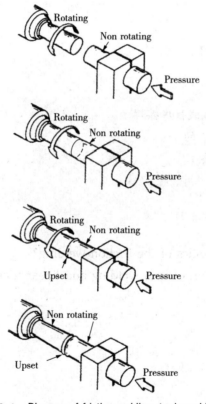

Fig. 30.1　Diagram of friction welding (spin welding)

30.3.2　Linear friction welding

Linear friction welding (LFW) is similar to spin welding except that the moving chuck oscillates laterally instead of spinning. The speeds are much lower in general, which requires the pieces to be kept under pressure at all times. This also requires the parts to have a high shear strength. Linear friction welding requires more complex machinery than spin welding, but has the advantage that parts of any shape can be joined, as opposed to parts with a circular meeting point. Another advantage is that in many instances quality of joint is better than that obtained using rotating technique.

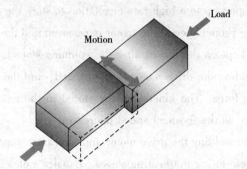

Fig. 30. 2　Diagram of friction welding (linear welding)

Words and terms

friction welding　摩擦焊

flywheel　飞轮

chuck　抛掷,丢弃

linear friction welding　线性摩擦焊

stationary　固定的,静止的

spin welding　旋转焊接

disengage　脱离,松开

Questions

What are the characteristics of the friction welding?

Please list the application of spin welding and linear friction welding, respectively.

31

Friction stir welding

31.1 Overview

Friction stir welding (FSW) is a solid-state joining process that uses a non-consumable tool to join two facing workpieces without melting the workpiece material, which is illustrated in Fig. 31.1. Heat is generated by friction between the rotating tool and the workpiece material, which leads to a softened region near the FSW tool. While the tool is traversed along the joint line, it mechanically intermixes the two pieces of metal, and forges the hot and softened metal by the mechanical pressure, which is applied by the tool, much like joining clay, or dough. It is primarily used on wrought or extruded aluminum and particularly for structures which need very high weld strength.

It was invented and experimentally proven at The Welding Institute (TWI) in the UK in December 1991. TWI held patents on the process, the first being the most descriptive.

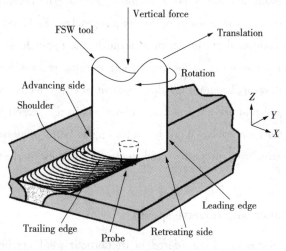

Fig. 31.1 Diagram of friction stir welding

31.2　Key parameters

31.2.1　Tool design

The design of the tool is a critical factor as a good tool can improve both the quality of the weld and the maximum possible welding speed.

It is desirable that the tool material be sufficiently strong, tough, and hard wearing at the welding temperature. Further it should have a good oxidation resistance and a low thermal conductivity to minimize heat loss and thermal damage to the machinery further up the drive train. Hot-worked tool steel such as AISI H13 has proven perfectly acceptable for welding aluminum alloys within thickness ranges of 0.5mm to 50 mm but more advanced tool materials are necessary for more demanding applications such as highly abrasive metal matrix composites or higher melting point materials such as steel or titanium.

Improvements in tool design have been shown to cause substantial improvements in productivity and quality. TWI has developed tools specifically designed to increase the penetration depth and thus increasing the plate thicknesses that can be successfully welded. An example is the "whorl" design that uses a tapered pin with re-entrant features or a variable pitch thread to improve the downwards flow of material. Additional designs include the Triflute and Trivex series. The Triflute design has a complex system of three tapering, threaded re-entrant flutes that appear to increase material movement around the tool. The Trivex tools use a simpler, non-cylindrical, pin and have been found to reduce the forces acting on the tool during welding.

The majority of tools have a concave shoulder profile which acts as an escape volume for the material displaced by the pin, prevents material from extruding out of the sides of the shoulder and maintains downwards pressure and hence good forging of the material behind the tool. The Triflute tool uses an alternative system with a series of concentric grooves machined into the surface which are intended to produce additional movement of material in the upper layers of the weld.

Widespread commercial applications of friction stir welding process for steels and other hard alloys such as titanium alloys will require the development of cost-effective and durable tools. Material selection, design and cost are important considerations in the search for commercially useful tools for the welding of hard materials. Work is continuing to better understand the effects of tool material's composition, structure, properties and geometry on their performance, durability and cost.

31.2.2　Tool rotation and traverse speeds

There are two tool speeds to be considered in friction-stir welding; how fast the tool rotates and how quickly it traverses along the interface. These two parameters have considerable importance and must be chosen with care to ensure a successful and efficient welding cycle. The relationship between the rotation speed, the welding speed and the heat input during welding is complex but, in general, it can be said that increasing the rotation speed or decreasing the traverse speed will result in a hotter weld. In order to produce a successful weld it is necessary that the material surrounding the tool is hot enough to enable the extensive plastic flow required and minimize the forces acting on the tool. If the material is too cold then voids or other flaws may be present in the stir zone and in extreme cases the tool may break.

Excessively high heat input, on the other hand may be detrimental to the final properties of the weld. Theoretically, this could even result in defects due to the liquation of low-melting-point phases (similar to liquation cracking in fusion welds). These competing demands lead onto the concept of a "processing window": the range of processing parameters viz. tool rotation and traverse speed, that will produce a good quality weld. Within this window the resulting weld will have a sufficiently high heat input to ensure adequate material plasticity but not so high that the weld properties are excessively deteriorated.

31.2.3　Tool tilt and plunge depth

The plunge depth is defined as the depth of the lowest point of the shoulder below the surface of the welded plate and has been found to be a critical parameter for ensuring weld quality. Plunging the shoulder below the plate surface increases the pressure below the tool and helps ensure adequate forging of the material at the rear of the tool. Tilting the tool by 2° to 4°, such that the rear of the tool is lower than the front, has been found to assist this forging process. The plunge depth needs to be correctly set, both to ensure the necessary downward pressure is achieved and to ensure that the tool fully penetrates the weld. Given the high loads required, the welding machine may deflect and so reduce the plunge depth compared to the nominal setting, which may result in flaws in the weld. On the other hand, an excessive plunge depth may result in the pin rubbing on the backing plate surface or a significant undermatch of the weld thickness compared to the base material. Variable load welders have been developed to automatically compensate for changes in the tool displacement while TWI has demonstrated a roller system that maintains the tool position above the weld plate.

31.2.4　Welding forces

During welding a number of forces will act on the tool:

A downwards force is necessary to maintain the position of the tool at or below the material surface. Some friction-stir welding machines operate under load control but in many cases the vertical position of the tool is preset and so the load will vary during welding.

The traverse force acts parallel to the tool motion and is positive in the traverse direction. Since this force arises as a result of the resistance of the material to the motion of the tool it might be expected that this force will decrease as the temperature of the material around the tool is increased.

The lateral force may act perpendicular to the tool traverse direction and is defined here as positive towards the advancing side of the weld.

Torque is required to rotate the tool, the amount of which will depend on the down force and friction coefficient (sliding friction) and/or the flow strength of the material in the surrounding region (stiction).

In order to prevent tool fracture and to minimize excessive wear and tear on the tool and associated machinery, the welding cycle is modified so that the forces acting on the tool are as low as possible, and abrupt changes are avoided. In order to find the best combination of welding parameters, it is likely that a compromise must be reached, since the conditions that favour low forces (e. g. high heat input, low travel speeds) may be undesirable from the point of view of productivity and weld properties.

Words and terms

    intermix   混合,混杂

    descriptive   描写的

    concave   凹面的

    flaw   缺陷

    lateral force   横向力

    whorl   盘旋,螺纹

    solid-state welding   固相焊接

    plunge depth   插入深度

Questions

The cost of friction stir welding is relatively high, please discuss why.

Why the welding tool needs to be tilted?

31.3　Flow of material

Early work on the mode of material flow around the tool used inserts of a different alloy, which had a different contrast to the normal material when viewed through a microscope, in an effort to determine where material was moved as the tool passed. The data was interpreted as representing a form of in-situ extrusion where the tool, backing plate and cold base material form the "extrusion chamber" through which the hot, plasticised material is forced. In this model the rotation of the tool draws little or no material around the front of the probe instead the material parts in front of the pin and passes down either side. After the material has passed the probe the side pressure exerted by the "die" forces the material back together and consolidation of the join occurs as the rear of the tool shoulder passes overhead and the large down force forges the material.

More recently, an alternative theory has been advanced that advocates considerable material movement in certain locations. This theory holds that some material does rotate around the probe, for at least one rotation, and it is this material movement that produces the "onion-ring" structure in the stir zone. The researchers used a combination of thin copper strip inserts and a "frozen pin" technique, where the tool is rapidly stopped in place. They suggested that material motion occurs by two processes:

1) Material on the advancing side of a weld enters into a zone that rotates and advances with the profiled probe. This material was very highly deformed and sloughs off behind the pin to form arc-shaped features when viewed from above (i. e. down the tool axis). It was noted that the copper entered the rotational zone around the pin, where it was broken up into fragments. These fragments were only found in the arc shaped features of material behind the tool.

2) The lighter material came from the retreating side in front of the pin and was dragged around to the rear of the tool and filled in the gaps between the arcs of advancing side material. This material did not rotate around the pin and the lower level of deformation resulted in a larger grain size.

The primary advantage of this explanation is that it provides a plausible explanation for the production of the onion-ring structure.

The marker technique for friction stir welding provides data on the initial and final positions of the marker in the welded material. The flow of material is then reconstructed from these positions. Detailed material flow field during friction stir welding can also be calculated from theoretical considerations based on fundamental scientific principles. Material flow calculations are routinely used in numerous engineering applications. Calculation of material flow fields in friction stir welding can be undertaken both using comprehensive numerical simulations or simple but insightful analytical equations. The comprehensive models for the calculation of material flow fields also provide important information such as geometry of the stir zone and the torque on the tool. The numerical simulations have shown the ability to correctly predict the results from marker experiments and the stir zone geometry observed in friction stir welding experiments.

31.4 Applications

The FSW process has initially been patented by TWI in most industrialised countries and licensed for over 183 users. Friction stir welding and its variants friction stir spot welding and friction stir processing are used for the following industrial applications：shipbuilding and offshore, aerospace, automotive, rolling stock for railways, general fabrication, robotics, and computers.

31.4.1　Shipbuilding and offshore

Two Scandinavian aluminium extrusion companies were the first to apply FSW commercially to the manufacture of fish freezer panels at Sapa in 1996, as well as deck panels and helicopter landing platforms at Marine Aluminium Aanensen. Marine Aluminium Aanensen subsequently merged with Hydro Aluminium Maritime to become Hydro Marine Aluminium. Some of these freezer panels are now produced by Riftec and Bayards. In 1997 two-dimensional friction stir welds in the hydro-dynamically flared bow section of the hull of the ocean viewer vessel. The Boss were produced at Research Foundation Institute with the first portable FSW machine. The Super Liner Ogasawara at Mitsui Engineering and Shipbuilding is the largest friction stir welded ship so far. The Sea Fighter of Nichols Bros and the Freedom class Littoral Combat Ships contain prefabricated panels by the FSW fabricators Advanced Technology and Friction Stir Link, Inc. respectively. The Houbei class missile boat has friction stir welded rocket launch containers of China Friction Stir Centre. HMNZS Rotoiti in New Zealand has FSW panels made by Donovans in a converted milling machine. Various companies apply FSW to armor plating for amphibious assault ships.

31.4.2　Aerospace

United Launch Alliance applies FSW to the Delta Ⅱ, Delta Ⅳ, and Atlas V expendable launch vehicles, and the first of these with a friction stir welded interstage module was launched in 1999. The process is also used for the Space Shuttle external tank, for Ares I and for the Orion Crew Vehicle test article at NASA as well as Falcon 1 and Falcon 9 rockets at SpaceX. The toe nails for ramp of Boeing C-17 Globemaster Ⅲ cargo aircraft by Advanced Joining Technologies and the cargo barrier beams for the Boeing 747 Large Cargo Freighter were the first commercially produced aircraft parts. FAA approved wings and fuselage panels of the Eclipse 500 aircraft were made at Eclipse Aviation, and this company delivered 259 friction stir welded business jets, before they were forced into Chapter 7 liquidation. Floor panels for Airbus A400M military aircraft are now made by Pfalz Flugzeugwerke and Embraer used FSW for the Legacy 450 and 500 Jets Friction stir welding also is employed for fuselage panels on the Airbus A380. BRÖTJE-Automation uses friction stir welding for gantry production machines developed for the aerospace sector as well as other industrial applications.

31.4.3　Automotive

Aluminium engine cradles and suspension struts for stretched Lincoln Town Car were the first automotive parts that were friction stir at Tower Automotive, who use the process also for the engine tunnel of the Ford GT. A spin-off of this company is called Friction Stir Link, Inc. and successfully exploits the FSW process, e. g. for the flatbed trailer "Revolution" of Fontaine Trailers. In Japan FSW is applied to suspension struts at Showa Denko and for joining of aluminium sheets to galvanized steel brackets for the boot (trunk) lid of the Mazda MX-5. Friction stir spot welding is successfully used for the bonnet (hood) and rear doors of the Mazda RX-8 and the boot lid of the Toyota Prius. Wheels are friction stir welded at Simmons Wheels, UT Alloy Works and Fundo. Rear seats for the Volvo V70 are friction stir welded at Sapa, HVAC pistons at Halla Climate Control and exhaust gas recirculation coolers at Pierburg. Tailor welded blanks are friction stir welded for the Audi R8 at Riftec. The B-column of the Audi R8 Spider is friction stir welded from two extrusions at Hammerer Aluminium Industries in Austria.

31.4.4　Railways

Since 1997 roof panels were made from aluminium extrusions at Hydro Marine Aluminium with a bespoke 25 m long FSW machine, e. g. for DSB class SA-SD trains of Alstom LHB Curved side and roof panels for the Victoria line trains of London Underground, side panels for Bombardier's Electrostar trains at Sapa Group and side panels for Alstom's British Rail Class 390 Pendolino trains are made at Sapa Group Japanese commuter and express A-trains, and British Rail Class 395 trains are friction stir welded by Hitachi, while Kawasaki applies friction stir spot welding to roof panels and Sumitomo Light Metal produces Shinkansen floor panels. Innovative FSW floor panels are made by Hammerer Aluminium Industries in Austria for the Stadler KISS double decker rail cars, to obtain an internal height of 2 m on both floors and for the new car bodies of the Wuppertal Suspension Railway.

Heat sinks for cooling high-power electronics of locomotives are made at Sykatek, EBG, Austerlitz Electronics, Euro Composite, Sapa and Rapid Technic, and are the most common application of FSW due to the excellent heat transfer.

31.5　Advantages and disadvantages

Advantages:

Can produce high quality welds in a short cycle time;

No filler metal is required and flux is not used;

The process is capable of welding most of the common metals. It can also be used to join many combinations of dissimilar metals. Friction welding requires relatively expensive apparatus similar to a machine tool;

Easy to operate equipment;

Not time consuming;

Low levels of oxide films and surface impurities;

When compared to resistance butt welding creates better welds at lower cost and higher speed, lower levels of electric current are required;

Small heat affected zone when comparing the process to conventional flash welding;

When compared to flash butt welding, less shortening of the component;

No need to use gas, filler metal or flux. No slag that can cause weld imperfections.

Disadvantages:

Process limited to angular and flat butt welds;

Only used for smaller parts;

Complicated when used for tube welding;

Hard to remove flash when working with high carbon steel;

Requires a heavy rigid machine in order to create high thrust pressure.

Words and terms

onion-ring 洋葱环

theoretical 理论的,理论上的

insightful 有深刻了解的

offshore 近海的

military 军事的

thrust pressure 推力

plausible explanation 合理的解释

numerical simulation 数值模拟

analytical equation 分析方程

amphibious 两栖的,水陆两用的

flash welding 闪光焊

Questions

What factors can affect the grain size of the friction stir welding microstructure?

What do you think is the most notable advantage of friction stir welding?

32

Explosive welding

32.1 Overview

Explosion welding (EXW) is a solid state (solid-phase) process where welding is accomplished by accelerating one of the components at extremely high velocity through the use of chemical explosives, as illustrated in Fig. 32.1. This process is most commonly utilized to clad carbon steel plate with a thin layer of corrosion resistant material (e. g., stainless steel, nickel alloy, titanium, or zirconium). Due to the nature of this process, producible geometries are very limited. They must be simple. Typical geometries produced include plates, tubing and tubesheets.

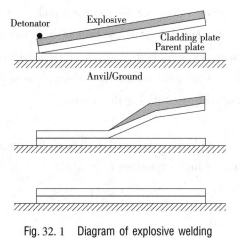

Fig. 32.1 Diagram of explosive welding

32.2 Development

Unlike other forms of welding such as arc welding (which was developed in the late 19th century), explosion welding was developed relatively recently, in the decades after World War Ⅱ. Its origins, however, go back to World War Ⅰ, when it was observed that pieces of shrapnel sticking to armor plating were not only embedding themselves, but were actually being welded to the metal. Since the extreme heat involved in other forms of welding did not play a role, it was concluded that the phenomenon was caused by the explosive forces acting on the shrapnel. These results were later duplicated in laboratory tests and, not long afterwards, the process was patented and put to use.

In 1962, DuPont applied for a patent on the explosion welding process, which was granted on June 23, 1964 under US Patent 3,137,937 and resulted in the use of the Detaclad trademark to describe the process. On July 22, 1996, Dynamic Materials Corporation completed the acquisition of DuPont's Detaclad operations for a purchase price of $ 5,321,850.

Recently, the response of inhomogeneous plates undergoing explosive welding was analytically modeled.

32.3 Advantages and disadvantages

Advantages:

Explosion welding can produce a bond between two metals that cannot necessarily be welded by conventional means. The process does not melt either metal, instead plasticizing the surfaces of both metals, causing them to come into intimate contact sufficient to create a weld. This is a similar principle to other non-fusion welding techniques, such as friction welding. Large areas can be bonded extremely quickly and the weld itself is very clean, due to the fact that the surface material of both metals is violently expelled during the reaction.

Disadvantages:

Extensive knowledge of explosives is needed before the procedure may be attempted safely. Regulations for the use of high explosives may require special licensing.

Words and terms

electroslag welding 电渣焊
laser beam welding 激光焊
explosive welding 爆炸焊

clad　覆盖

duplicated　复制出的

electron beam welding　电子束焊

friction stir welding　搅拌摩擦焊

zirconium　锆

shrapnel　弹片,零钱

acquisition　获得,收购

Questions

When is the explosive welding preferred?

What are the most important factors that affect the bonding quality of explosive welding?

33

Brazing

33. 1 Overview

Brazing is a group of welding processes which produces coalescence of materials by heating to a suitable temperature and using a filler metal having a liquidus above 840 °F (449 ℃) and below the solidus of the base metals. The filler metal is distributed between the closely fitted surfaces of the joint by capillary attraction. Brazing is distinguished from soldering in that soldering employs a filler metal having a liquidus below 840 °F (449 ℃).

When brazing with silver alloy filler metals (silver soldering), the alloys have liquidus temperatures above 840 °F (449 ℃).

Brazing is only used on ferrous metals because the solder melts at 960 ℃, a point above the melting point of non-ferrous metals.

Brazing must meet each of three criteria:

1) The parts must be joined without melting the base metals.

2) The filler metal must have a liquidus temperature above 840 °F (449 ℃).

3) The filler metal must wet the base metal surfaces and be drawn onto or held in the joint by capillary attraction.

33. 2 Base brazing metals

In addition to the normal mechanical requirements of the base metal in the brazement, the effect of the brazing cycle on the base metal and the final joint strength must be considered. Cold-work strengthened base metals will be annealed when the brazing process temperature and time are

192

in the annealing range of the base metal being processed. "Hot-cold worked" heat resistant base metals can also be brazed; however, only the annealed physical properties will be available in the brazement. The brazing cycle will usually anneal the cold worked base metal unless the brazing temperature is very low and the time at heat is very short. It is not practical to cold work the base metal after the brazing operation.

When a brazement must have strength above the annealed properties of the base metal after the brazing operation, a heat treatable base metal should be selected. The base metal can be an oil quench type, an air quench type that can be brazed and hardened in the same or separate operation, or a precipitation hardening type in which the brazing cycle and solution treatment cycle may be combined. Hardened parts may be brazed with a low temperature filler metal using short times at temperature to maintain the mechanical properties.

The strength of the base metal has an effect on the strength of the brazed joint. Some base metals are also easier to braze than others, particularly by specific brazing processes. For example, a nickel base metal containing high titanium or aluminum additions will present special problems in furnace brazing. Nickel plating is sometimes used as a barrier coating to prevent the oxidation of the titanium or aluminum, and it presents a readily wettable surface to the brazing filler metal.

33.3 Brazing filler metals

For satisfactory use in brazing applications, brazing filler metals must possess the following properties:

1) The ability to form brazed joints possessing suitable mechanical and physical properties for the intended service application.

2) A melting point or melting range compatible with the base metals being joined and sufficient fluidity at brazing temperature to flow and distribute into properly prepared joints by capillary action.

3) A composition of sufficient homogeneity and stability to minimize separation of constituents (liquation) under the brazing conditions to be encountered.

4) The ability to wet the surfaces of the base metals being joined and form a strong, sound bond.

5) Depending on the requirements, ability to produce or avoid base metal-filler metal interactions.

33.4 **Principles**

Capillary flow is the most important physical principle which ensures good brazements providing both adjoining surfaces molten filler metal. The joint must also be properly spaced to permit efficient capillary action and resulting coalescence. More specifically, capillarity is a result of surface tension between base metal(s), filler metal, flux or atmosphere, and the contact angle between base and filler metals. In actual practice, brazing filler metal flow characteristics are also influenced by considerations involving fluidity, viscosity, vapor pressure, gravity, and by the effects of any metallurgical reactions between the filler and base metals.

The brazed joint, in general, is one of a relatively large area and very small thickness. In the simplest application of the process, the surfaces to be joined are cleaned to remove contaminants and oxide. Next, they are coated with flux or a material capable of dissolving solid metal oxides present and preventing new oxidation. The joint area is then heated until the flux melts and cleans the base metals, which are protected against further oxidation by the liquid flux layer.

Brazing filler metal is then melted at some point on the surface of the joint area. Capillary attraction is much higher between the base and filler metals than that between the base metal and flux. Therefore, the flux is removed by the filler metal. The joint, upon cooling to room temperature, will be filled with solid filler metal. The solid flux will be found on the joint surface.

High fluidity is a desirable characteristic of brazing filler metal because capillary attraction may be insufficient to cause a viscous filler metal to run into tight fitting joints.

Brazing is sometimes done with an active gas, such as hydrogen, or in an inert gas or vacuum. Atmosphere brazing eliminates the necessity for post cleaning and ensures absence of corrosive mineral flux residue. Carbon steels, stainless steels, and super alloy components are widely processed in atmospheres of reacted gases, dry hydrogen, dissociated ammonia, argon, and vacuum. Large vacuum furnaces are used to braze zirconium, titanium, stainless steels, and the refractory metals. With good processing procedures, aluminum alloys can also be vacuum furnace brazed with excellent results.

Brazing is a process preferred for making high strength metallurgical bonds and preserving needed base metal properties because it is economical.

Smooth Joints Created Using Fillet Brazing over Acetylene Welding for Bicycle Frame.

33.5 **Tips**

The two metals being joined need to have a close and optimal fit in order for the capillary action of the brazing alloy filler metal to be drawn in.

Contaminants on the metal surface can inhibit the capillary process for the filler metal. This will reduce the strength of the joint being brazed. Chemical and mechanical cleaning processes can be used.

Chemical

- petroleum spray
- chlorinated solvents
- vapor de-greasing
- emulsion spray
- alkaline soak
- acid pickling
- trichlor solvents

Mechanical

- sandblasting (exercise caution to avoid damage to metal surface)
- machining
- grinding
- brushing (wire brush)

Make the minimally necessary use of vises or clamps needed to enable the joint to be self-supporting (if possible). Avoid or at least be aware of distortions caused by heat absorbing material such as the clamps.

Flux helps with oxidation problems caused by oxygen in the air and gas. Too much oxidation will interfere with the capillary effect of the filler alloy. Some filler metals already contain agents that act as flux.

The brazing method selected depends on the type and size of the job to be performed. For smaller jobs oxy-acetylene torch brazing is a common approach. In other types of jobs, the brazing processes such as resistance, induction, vacuum and atmosphere furnace brazing can be more efficient. One tip when brazing is to realize that filler metals attract to the highest temperature surface. Therefore if you heat directly on the joint surface the brazing alloy may not fill the joint. Instead, the goal is to heat the interior facing surfaces to the correct temperature, and by situating the brazing alloy close to the joint to be brazed.

Residues can often be removed with a hot water bath after the filler has solidified. If this doesn't crack the residue off, try a water jet and wire brush. As a last resort a mild acid bath. Follow the manufacturer's directions in order to avoid acid etching on the brazed metal.

Simplify the types of brazing rods needed by using a versatile product such as the HTS-2000. It works on all non-ferrous alloys including "all" aluminum alloys (even the ones that cannot be welded), magnesium aluminum mixtures, zinc, die cast, pot metal, copper, bronze, Nickel, Titanium and galvanized parts.

Words and terms

liquidus　液体的；液相线

brazing　钎焊

viscous filler　黏滞产品灌装机

petroleum　石油；原油

vapor de-greasing　蒸汽除油

alkaline soak　碱性浸渍

capillary attraction　毛细作用

fluidity　流动性

dissociated ammonia　解离氨

chlorinated solvents　氯化溶剂类

emulsion spray　乳化液喷雾

acid pickling　酸洗

trichlor solvents　九氯溶剂

sandblasting　喷砂处理

Questions

When is brazing preferred rather than other welding method?

Try to list three kinds of base materials and corresponding filler metals for brazing.

33.6　Processes

Generally, brazing processes are specified according to heating methods (sources) of industrial significance. Whatever the process used, the filler metal has a melting point above 840 ℉ (450 ℃) but below the base metal and distributed in the joint by capillary attraction.

33.6.1　Torch Brazing

Torch brazing is commonly used for smaller production runs or one assembly. This type of brazing is performed by heating with a gas torch set to the proper required composition, and an appropriate flux. This depends on the temperature and heat amount required. The fuel gas (acetylene, propane, city gas, etc.) may be burned with air, compressed air, or oxygen.

Brazing filler metal may be pre-placed at the joint in the forms of rings, washers, strips, slugs, or powder, or it may be fed from hand-held filler metal in wire or rod form. In any case, proper cleaning and fluxing are essential.

For manual torch brazing, the torch may be equipped with a single tip, either single or multiple flame. Manual torch brazing is particularly useful on assemblies involving sections of unequal mass. Welding machine operations can be set up where the production rate allows, using one or several torches equipped with single or multiple flame tips. The machine may be designed to move either the work or torches, or both. For premixed city gas-air flames, a refractory type burner is used.

33.6.2 Furnace brazing

Furnace brazing is used extensively where the parts to be brazed can be assembled with the brazing filler metal in form of wire, foil, filings, slugs, powder, paste, or tape is pre-placed near or in the joint. This process is particularly applicable for high production brazing. Fluxing is employed except when an atmosphere is specifically introduced in the furnace to perform the same function. Most of the high production brazing is done in a reducing gas atmosphere, such as hydrogen and combusted gases that are either exothermic (formed with heat evolution) or endothermic (formed with heat absorption). Pure inert gases, such as argon or helium, are used to obtain special atmospheric properties.

A large volume of furnace brazing is performed in a vacuum, which prevents oxidation and often eliminates the need for flux. Vacuum brazing is widely used in the aerospace and nuclear fields, where reactive metals are joined or where entrapped fluxes would be intolerable. If the vacuum is maintained by continuous pumping, it will remove volatile constituents liberated during brazing. There are several base metals and filler metals that should not be brazed in a vacuum because low boiling point or high vapor pressure constituents may be lost. The types of furnaces generally used are either batch or contiguous. These furnaces are usually heated by electrical resistance elements, gas or oil, and should have automatic time and temperature controls. Cooling is sometimes accomplished by cooling chambers, which either are placed over the hot retort or are an integral part of the furnace design. Forced atmosphere injection is another method of cooling. Parts may be placed in the furnace singly, in batches, or on a continuous conveyor.

A vacuum is a relatively economical method of providing an accurately controlled brazing atmosphere. Vacuum provides the surface cleanliness needed for good wetting and flow of filler metals without the use of fluxes. Base metals containing chromium and silicon can be easily vacuum brazed where a very pure, low dew point atmosphere gas would otherwise be required.

33.6.3 Induction Brazing

In this process, the heat necessary to braze metals is obtained from a high frequency electric current consisting of a motor-generator, resonant spark gap, and vacuum tube oscillator. It is induced or produced without magnetic or electric contact in the parts (metals). The parts are placed in or near a water-cooled coil carrying alternating current. They do not form any part of the electrical circuit. The brazing filler metal normally is pre-placed.

Careful design of the joint and the coil setup are necessary to assure that the surfaces of all members of the joint reach the brazing temperature at the same time.

Flux is employed except when an atmosphere is specifically introduced to perform the same function.

The equipment consists of tongs or clamps with the electrodes attached at the end of each arm. The tongs should preferably be water-cooled to avoid overheating. The arms are current carrying conductors attached by leads to a transformer. Direct current may be used but is comparatively expensive. Resistance welding machines are also used. The electrodes may be carbon, graphite, refractory metals, or copper alloys according to the required conductivity.

33.6.4　Resistance

The heat necessary for resistance brazing is obtained from the resistance to the flow of an electric current through the electrodes and the joint to be brazed. The parts comprising the joint form a part of the electric circuit. The brazing filler metal, in some convenient form, is pre-placed or face fed. Fluxing is done with due attention to the conductivity of the fluxes. (Most fluxes are insulators when dry.) Flux is employed except when an atmosphere is specifically introduced to perform the same function. The parts to be brazed are held between two electrodes, and proper pressure and current are applied. The pressure should be maintained until the joint has solidified. In some cases, both electrodes may be located on the same side of the joint with a suitable backing to maintain the required pressure.

33.6.5　Dip brazing

There are two methods of dip brazing: chemical bath dip brazing, molten metal bath dip brazing.

- **Chemical bath dip brazing**

In chemical bath dip brazing, the brazing filler metal, in suitable form, is pre-placed and the assembly is immersed in a bath of molten salt. The salt bath furnishes the heat necessary for brazing and usually provides the necessary protection from oxidation; if not, a suitable flux should be used. The salt bath is contained in a metal or other suitable pot, also called the furnace, which is heated from the outside through the wall of the pot, by means of electrical resistance units placed in the bath, or by the I2R loss in the bath itself.

- **Molten metal bath dip brazing**

In molten metal bath dip brazing, the parts are immersed in a bath of molten brazing filler metal contained in a suitable pot. The parts must be cleaned and fluxed if necessary. A cover of flux should be maintained over the molten bath to protect it from oxidation. This method is largely confined to brazing small parts, such as wires or narrow strips of metal. The ends of the wires or parts must be held firmly together when they are removed from the bath until the brazing filler metal has fully solidified.

33.6.6　Infrared brazing

Infrared heat is radiant heat obtained below the red rays in the spectrum. While with every "black" source there is sane visible light, the principal heating is done by the invisible radiation. Heat sources (lamps) capable of delivering up to 5,000 watts of radiant energy are commercially available. The lamps do not necessarily need to follow the contour of the part to be heated even though the heat input varies inversely as the square of the distance from the source. Reflectors are used to concentrate the heat.

Assemblies to be brazed are supported in a position that enables the energy to impinge on the part. In some applications, only the assembly itself is enclosed. There are, however, applications where the assembly and the lamps are placed in a bell jar or retort that can be evacuated, or in which an inert gas atmosphere can be maintained. The assembly is then heated to a controlled temperature, as indicated by thermocouples. The part is moved to the cooling platens after brazing.

Blanket brazing

Blanket brazing is another of the processes used for brazing. A blanket is resistance heated, and most of the heat is transferred to the parts by two methods, conduction and radiation, the latter being responsible for the majority of the heat transfer.

Exothermic brazing

Exothermic brazing is another special process by which the heat required to melt and flow a commercial filler metal is generated by a solid state exothermic chemical reaction. An exothermic chemical reaction is defined as any reaction between two or more reactants in which heat is given off due to the free energy of the system. Nature has provided us with countless numbers of these reactions; however, only the solid state or nearly solid state metal-metal oxide reactions are suitable for use in exothermic brazing units. Exothermic brazing utilizes simplified tooling and equipment.

The process employs the reaction heat in bringing adjoining or nearby metal interfaces to a temperature where pre-placed brazing filler metal will melt and wet the metal interface surfaces. The brazing filler metal can be a commercially available one having suitable melting and flow temperatures. The only limitations may be the thickness of the metal that must be heated through and the effects of this heat, or any previous heat treatment, on the metal properties.

Words and terms

torch brazing　火焰钎焊
endothermic　吸热的
furnace brazing　炉内钎焊

volatile constituent　挥发性组分

continuous conveyor　连续输送机

exothermic brazing　放热钎焊

resonant spark gap　谐振火花隙

molten metal bath dip brazing　熔化金属浸钎焊

chemical bath dip brazing　化学浸钎焊

vacuum　真空

infrared brazing　红外线钎焊

blanket brazing　保护硬杆焊

Questions

Please briefly describe the main characteristics of each brazing type.

Which brazing type do you think is the most practical one? Why?

34
Soldering

34. 1　**Overview**

Soldering is a group of processes that join metals by heating them to a suitable temperature. A filler nonferrous metal that melts at a temperature below 840 ℉ (449 ℃) and below that of the metals to be joined is used. The filler metal is distributed between the closely fitted surfaces of the joint by capillary attraction. Soldering uses fusible alloys to join metals (brazing occurs at temperatures above 840 ℉).

34. 2　**Types of soldering**

- Torch soldering: soldering process using air-fuel or oxy-fuel torches. Application can be automatic or manual.
- Furnace: parts are soldered by passing them through a furnace.
- Iron
- Induction
- Resistance
- Dip (small scale process for electronic components)
- Infrared
- Ultrasonic
- Reflow or paste
- Wave (used to attach circuits to circuit boards)

34.3　Soldering tips

There are three types of soldering tool tips, as shown in Fig. 34.1.

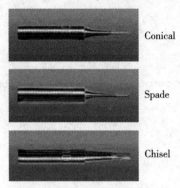

Conical

Spade

Chisel

Fig. 34.1　Typical soldering tips

34.4　Solder types

- **Hard soldering**

Hard soldering or silver soldering refers to solder that has silver content used to lower the melting point so that the molten metal flows more easily. This type of soldering requires a hot heat source, requiring a special torch. Oxyacetylene equipment can also be used, but there is a risk that some metals can be melted such as copper. Hard soldering is considered to be one of the best methods for joining two copper parts.

- **Soft soldering**

This process is used for joining most common metals with an alloy that melts at a temperature below that of the base metal. In many respects, this operation is similar to brazing in that the base is not melted, but is merely tinned on the surface by the solder filler metal. For its strength the soldered joint depends on the penetration of the solder into the pores of the base metal surface, along with the consequent formation of a base metal-solder alloy, together with the mechanical bond between the parts. Soft solders are used for airtight or watertight joints which are not exposed to high temperatures.

34.5　Application of solder

Soft solder joints may be made by using gas flames, wiping, sweating the joints, or by dipping in solder baths. Dipping is particularly applicable to the repair of radiator cores. Electrical connections and sheet metal are soldered with a soldering iron or gun. Wiping is a method used for joining lead pipe and also the lead jacket of underground and other lead-covered cables. Sweated joints may be made by applying a mixture of solder powder and paste flux to the joints. Then heat the part until this solder mixture liquifies and flows into the joints, or tin mating surfaces of members to be joined, and apply heat to complete the joint.

34.7 Differences between brazing vs welding and soldering

Brazing is not the same as braze welding.

Brazing is capillary, with heat drawing a filler metal using a capillary action. Capillary action refers to the way filler metal is drawn into a properly fitted joint. A simple example is the way water adheres to and is drawn up a straw in opposition to external forces such as gravity.

Braze welding bridges a gap using filler metal that is melted and deposited in fillets and grooves exactly at the points it is to be used. Limited base metal fusion may occur in braze welding.

Soldering is a process that joins materials with a filler metal that has a liquidus not greater than 842 °F (450 ℃). Like brazing, the filler metal moves into the joint with a capillary action.

To achieve a good joint using any of the various brazing processes, the parts must be properly cleaned and protected by either flux or the atmosphere during heating to prevent excessive oxidation. The parts must provide a capillary for the filler metal when properly aligned, and a heating process must be selected that will provide proper brazing temperatures and heat distribution.

Brazing vs. soldering

- Soldering: filler metals have a melting point below 840 °F (450 ℃)
- Brazing: filler metals have a melting point above 840 °F (450 ℃)

Welding refers to fusion of two metals, while soldering and brazing use adhesion.

In soldering and brazing the filler metal melts and flows into the joint. The base material remains intact or un-melted. The parts are fitted with tight tolerances, which produce a capillary action (capillarity) to draw the filler metal into the joint.

The advantages of brazing and soldering include:

- ability to join metals that cannot be welded. reheating can separate parts, particularly if one needs to be replaced
- easier to separate joined parts
- parts can be produced in a batch furnace

- portable process for joining smaller parts

The downside of both soldering and brazing include：

- tight joint tolerance required for capillary action
- lower strength vs. welding
- larger metal parts need to be soldered or brazed in a big furnace
- flux is required

Words and terms

brazing   钎焊

capillary action   毛细反应

dip brazing   浸钎焊

spatter   喷溅

disintegrate   使分解

conical   圆锥的

chisel   凿子

reflow   回流

disintegrate   使分解

soldering   低温钎焊

insulators   绝缘体

infrared   红外线

amplitude   振幅

encompass   包含

spade   铲子

induction   感应

insulators   绝缘子

encompass   包含,环绕,完成

Questions

What's the difference between brazing and soldering?

Name the common filler materials for brazing and soldering.

Talk about the application of brazing and soldering.

35

3D printing and additive manufacturing

35.1 Introduction

3D printing or additive manufacturing is the construction of a three-dimensional object from a CAD model or a digital 3D model. It can be done in a variety of processes in which material is deposited, joined or solidified under computer control, with material being added together, typically layer by layer.

Up to now, the precision, repeatability, and material range of 3D printing have increased to the point that some 3D printing processes are considered viable as an industrial-production technology, whereby the term additive manufacturing can be used synonymously with 3D printing. One of the key advantages of 3D printing is the ability to produce very complex shapes or geometries that would be otherwise impossible to construct by hand, including hollow parts or parts with internal truss structures to reduce weight. Fused deposition modeling (FDM), which uses a continuous filament of a thermoplastic material, is the most common 3D printing process in use as of 2020.

The umbrella term additive manufacturing (AM) gained popularity in the 2000s, inspired by the theme of material being added together. In contrast, the term subtractive manufacturing appeared as a retronym for the large family of machining processes with material removal as their common process. The term 3D printing still referred only to the polymer technologies in most minds, and the term AM was more likely to be used in metalworking and end-use part production contexts than among polymer, inkjet, or stereolithography enthusiasts.

35.2　Additive Manufacturing Technologies

According to the process characteristics of additive manufacturing, the current additive manufacturing technologies can be divided into the following seven categories: vat photopolymerization, material extrusion, material jetting, binder jetting, powder bed fusion, sheet lamination and directed energy deposition. Polymer can be fabricated via vat photopolymerization, material extrusion, material jetting, binder jetting and powder bed fusion. While metallic products can be manufactured by powder bed fusion, sheet lamination and directed energy deposition.

35.3　Modeling

3D printable models may be created with a computer-aided design (CAD) package, via a 3D scanner, or by a plain digital camera and photogrammetry software. 3D printed models created with CAD result in relatively fewer errors than other methods. Errors in 3D printable models can be identified and corrected before printing. The manual modeling process of preparing geometric data for 3D computer graphics is similar to plastic arts such as sculpting. 3D scanning is a process of collecting digital data on the shape and appearance of a real object, creating a digital model based on it.

CAD models can be saved in the stereolithography file format (STL), a de facto CAD file format for additive manufacturing that stores data based on triangulations of the surface of CAD models. STL is not tailored for additive manufacturing because it generates large file sizes of topology optimized parts and lattice structures due to the large number of surfaces involved. A newer CAD file format, the Additive Manufacturing File format (AMF) was introduced in 2011 to solve this problem. It stores information using curved triangulations.

35.4　Printing

Before printing a 3D model from an STL file, it must first be examined for errors. Most CAD applications produce errors in output STL files, such as holes, faces normal, self-intersections, noise shells, manifold errors, overhang issues.

A step in the STL generation known as "repair" fixes such problems in the original model. Generally STLs that have been produced from a model obtained through 3D scanning often have more of these errors as 3D scanning is often achieved by point to point acquisition/

mapping. 3D reconstruction often includes errors.

Once completed, the STL file needs to be processed by a piece of software called a "slicer", which converts the model into a series of thin layers and produces a G-code file containing instructions tailored to a specific type of 3D printer (FDM printers). This G-code file can then be printed with 3D printing client software (which loads the G-code, and uses it to instruct the 3D printer during the 3D printing process).

Printer resolution describes layer thickness and X-Y resolution in dots per inch (dpi) or micrometers (μm). Typical layer thickness is around 100 μm (250 DPI), although some machines can print layers as thin as 16 μm (1,600 DPI). X-Y resolution is comparable to that of laser printers. The particles (3D dots) are around 50 to 100 μm (510 to 250 DPI) in diameter. For that printer resolution, specifying a mesh resolution of 0.01-0.03 mm and a chord length$\leqslant$0.016 mm generates an optimal STL output file for a given model input file. Specifying higher resolution results in larger files without increase in print quality.

Construction of a model with contemporary methods can take anywhere from several hours to several days, depending on the method used and the size and complexity of the model. Additive systems can typically reduce this time to a few hours, although it varies widely depending on the type of machine used and the size and number of models being produced simultaneously.

35.5　Materials

Traditionally, 3D printing focused on polymers for printing, due to the ease of manufacturing and handling polymeric materials. However, the method has rapidly evolved to not only print various polymers but also metals and ceramics, making 3D printing a versatile option for manufacturing. 3D printing is presently available in manufacturing polymers, ceramics, structural materials, metamaterials, magnetic materials, piezoelectric materials, biological materials and smart materials.

35.6　Multi-material 3D printing

A drawback of many existing 3D printing technologies is that they only allow one material to be printed at a time, limiting many potential applications which require the integration of different materials in the same object. Multi-material 3D printing solves this problem by allowing objects of complex and heterogeneous arrangements of materials to be manufactured using a single printer.

Words and terms

additive manufacturing　增材制造

vat photopolymerization　光固化

Material Extrusion　材料挤出

Material Jetting　材料喷射

Binder Jettin　黏合剂喷射

Powder Bed Fusion　粉末床熔融

Sheet Lamination　薄材叠层

Directed Energy Deposition　定向能量沉积

Inkjet　喷墨

deposit　沉积

Questions

What is 3D printing?

What is the application of additive manufacturing?

第二部分 **专业英语翻译**

36

专业英语的特点概述

专门用途英语(English for Specific Purpose, ESP),又称专业英语,由世界著名语言学家 M. A. K. Halliday(1964)在 *The Linguistic Sciences and Language Teaching* 一书中首次提出,并在西方一些发达的国家掀起了教学和研究活动。在我国,自从 20 世纪 80 年代兴起科技英语热之后,ESP 也开始受到广泛关注。ESP 是随着科学技术的发展而形成的一种独立的文体形式,既涵盖自然科学领域的各种知识和技术,也包括社会科学的各个领域。如用英语撰写的有关自然科学和社会科学的学术著作、论文、技术标准、实验报告、专利产品的说明书、设备使用说明书、新闻、科普读物、科技报道等多种类型和文体。在科技和跨文化交流和合作中,专业技术人员除了掌握通用英语(English for General Purpose, EGP)以满足日常所需的听、说、写、读、译的基本能力,还需具备听懂和理解英语学术报告的能力、在国内外参加学术会议时用英语进行学术交流的能力、在各种专业研讨会上用英语参加辩论和答辩的能力、在阅读和听课时用英语快速记笔记的能力以及在长篇学术报告中用英语做摘录的能力等。

不同于通用英语,作为学习专业英语阅读、翻译和写作的第一步,本章介绍了专业英语在词汇、语法、语篇上的特点。

36.1 专业英语的语言特点概述

专业英语是传达专业知识和信息的文体,它与通用英语,如文学英语、日常英语等文体相比,在形式、结构以及语法方面并没有什么不同,但它有其特点。

英国文学家赫胥黎(Aldous Hnxley)认为:"把语言所说不清楚的,说清楚了,是科学;把语言所不能表达的,表达出来,是文学"。专业英语涉及科学技术,而科学技术主要是以客观世界万事万物为研究对象,反映的是事物发展的客观规律,追求的是"求真"。客观事物运用英语对之进行描述或论证时,必然要求语言运用者坚持客观公允的态度,力求准确地反映客观现实,唯有如此方能充分地体现科学精神、揭示科学真理。同时专业英语强调事实,文体以议论文和说明文为主,行文不追求辞藻华丽、文采飞扬,通常十分简约、精炼、言简意赅。文学语言多强调情感,追求的是"艺术",倾向于主观的描写,文学是"1+1>2"。下面以科学家和翻译家对诗圣杜甫《偶题》中"文章千古事,得失寸心知"一句为例,说明科学语言和文学语言的区别。杨振宁在《杨振宁文选》英文版序言中将以上两句诗句翻译为"A piece of literature is meant for the millennium. But its ups and downs are known already in the author's heart."而长期从事文学翻译的北京大学教授许渊冲建议译为"A poem may long, long remain. Who knows the poets' loss and gain!"杨振宁的翻译十分准确,但属于科学语言,许渊冲认为将"文章"译成"literature"不妥,因为杜甫没写过多少文章,所以应该译为"poem",另外前者将"千古"译成"millennium",而后者译成"long, long remain",这就是文学语言了。

专业英语与日常英语也有区别,专业语言的表达方式和日常生活的表达方式是不相同的。对下面两句话作一个比较,就可以看到两者的不同之处。在日常交流中,我们说:"发生偏差,可能是径向摩擦系数不均匀,对称性破坏了",或"低碳钢没有中碳钢那么硬,因为含有较少的碳"。而科技文献往往这样表述:"偏差可能是由于径向摩擦系数不均而打破了对称性所致"。"低碳钢不如中碳钢硬,因为其含碳量较低"。可见日常英语和科学英语在措辞和表达方式上存在较大差异,见表36.1。

表36.1　日常英语与科学英语的特点比较

日常英语	科学英语
1.通俗化:常用词汇用得多	1.专业化:专业术语用得多
2.多义性:一词多义,使用范围广	2.单义性:词义相对单一,因专业而不同
3.人称化:人称丰富,形式多样	3.物称化:多用物称,以示客观
4.多时性:描述生活,时态多样	4.现时性:叙述事实多用现在时
5.主动态:句子倾向于主动态	5.被动态:句子倾向于被动态
6.简单性:单句用得多,散句用得多	6.复杂性:复杂单句用得多,复杂句用得多
7.口语化:口语用得多,随意灵活	7.书面化:长句用得多,书卷意浓

摘自(黄忠廉、李亚舒,《科学翻译学》,2007:164)。

请看下面的例子:

Casting is one of the metal working techniques known to man. China made metal castings as early as 2000 B. C., and the process used then is not much different in principle from one used to-day. Foundry process consist of making molds, preparing and melting the metal, pouring the metal into the molds, and cleaning the castings. The product of the foundry is a casting, which may vary from a fraction of a kilogram to several hundred tons. It may also vary in composition as practically all metals and alloys can be cast. The metals most frequently cast are iron, steel, aluminum and so on. Of these, iron, because of its low melting point, low price and ease of control, is outstanding for its suitability for casting and used far more than the others. Casting is widely used method of producing metal products, particularly those which are intricate. Since molten materials will readily take the shape of the container into which they are poured, it nearly as easy to cast fairly complex shapes as to produce simple forms.

铸造是人类所掌握的最古老的金属加工技术之一,中国早在公元前2000年就已经把金属制成铸件,而所用的工艺从原理上和现今的工艺没有多大区别。铸造工艺包括制模、备料和金属熔炼,金属液浇注入模和铸件清砂。铸造的产品是铸件,铸件可能在零点几千克到几百吨范围变化,实际上所有金属在成分上也是变化的,而且合金也可以铸造。最常用来铸造的金属是铸铁、钢、铝等。这些金属中,铁由于其低熔点、低价格和易控制,因而其铸造性最好,用途比其他金属更多。铸造工艺是一种广泛应用生产金属制件的方法,实际上铸造工艺是复杂的。因为熔融的金属易充满铸型型腔,形状相当复杂,但和形状简单的零件一样容易铸造。

这段英语文献,作者介绍了铸造的历史、铸造工艺、铸造材料以及铸造的特点,为了描述客观事物的发展,解释工艺原理,全部采用非人称语气。整段文字语言规范、条理清楚、简洁流畅,并且使用的是正式的书面体,文中词语也几乎都是专业英语词语,所以这段文字可以说是典型的专业英语。

36.2 专业英语的语法特点

语法是学习语言的基础。专业英语是在通用英语的基础上发展起来的,因此,无论是在语法还是词汇方面与通用英语没有绝对的界限,但在语言发展的长期过程中,专业英语由于题材内容和使用方式的特殊性而形成了文体、词汇和语言规律方面的特点,与日常英语、文学英语、新闻、报刊等形成了明显的区别。其不同之处主要表现在语法方面,专业英语语法的特点主要表现在:时态、被动语态、语气、情态动词、后置定语、多重复句、名词化等几个方面。

36.2.1 被动语态

被动语态便于论述客观事实,而科技文献多数场合描述的是客观现象、实验或结果,不必指出行为动作的执行者,故被动语态常用于科技文献。如"一天测温四次",是谁去测无关紧要。专业英语中一般不用"We(或They或People)measure the temperature four times a day",而会用被动句"The temperature is measured four times a day"。即令有"执行者",但在专业文章中,大部分情况是既要强调"执行者",又要突出"承受者"。这样,用被动句来表达,就简洁得多。因为,在被动句中,承受者是主语,得到强调,承受者与by组成介词短语用作状语,比较显眼而得到突出。如"阀门动作由伺服电机控制",在专业英语中就不会用"We(或They,或People)use a servo-motor to control the valve's movement",也不会用"A servo-motor controls the movement of the valve",而会用被动句"The movement of the valve is controlled by a servo-motor"。所以,被动句在专业英语中用得特别多。

［1］Under no circumstances can energy be created or destroyed.

译文：在任何情况下，能量即不能创造，也不能消灭。

［2］Annealing at a low temperature may be used to eliminate the residual stresses produced during cold working without affecting the mechanical properties of the finish part.

译文：在低温下退火，可用于消除在冷加工过程中产生的残余应力，而不致影响成品零件的力学性能。

［3］A hybrid laser processing technique base on the idea of solid freeform fabrication using SLS, with the incorporation of solid foil materials, was hypothesized as a means of joining foil material into lattice structures.

译文：基于 SLS 快速成型的混合激光技术，结合固体薄膜材料的掺入，可以作为将箔材料加入晶格结构的一种手段。

［4］These basic ingredients must be combined in an orderly sequence for the production of a sound casting.

译文：为了生产出合格的铸件，这些最基本的要素必须按照一定的顺序组合起来。

［5］Three continuous cooling transformation(CCT) diagrams for S355J2 steel were employed to study the effect of CCT variations on calculated residual stresses.

译文：利用 3 个 S355J2 钢的连续冷却转变图研究了 CCT 变化对计算残余应力的影响。

［6］The effect of aging temperature and aging time to new-type heat-resisting aluminum alloy forging structure and properties are introduced.

译文：通过对新型铝合金时效处理后的组织结构进行分析，研究了不同的时效温度和时间对该铝合金力学性能的影响。

36.2.2　非谓语动词

非谓语动词(不定式、动名词、分词)具有使句子成分衔接紧密的功能，并能阐明句子中成分相互之间的内在联系，而且还有简化文章的作用。这正好能满足专业英语的要求，所以，非谓语动词在专业英语中被大量使用。

1)不定式结构

由于专业科技文献中所涉及的行为动作多数是"某次、具体、要做"的行为动作，而不定式恰有此功能，因此在专业英语文献中使用较频繁。不定式的构成一般为"to+动词原形"或"not to+动词原形"，它在句中可起到名词、形容词、副词等作用，即通常可作主语、宾语、表语、定语、状语、补足语等六大成分。例如：

［1］To develop materials suitable for spent-nuclear-fuel containers, the effect of forced cooling on mechanical properties and fracture toughness of heavy section ductile iron was investigated.

译文：为了开发适用于核乏燃料容器的材料，研究了强制冷却对厚大断面球墨铸铁力学性能和断裂韧性的影响。

［2］Fracture analysis was conducted *to study* the influence of vermicular and slightly irregular spheroidal graphite on the fracture behavior of heavy section ductile iron.

译文:研究了蠕变和少量不规则的球状石墨的断口分析对厚大断面球墨铸铁的断裂行的影响。

[3] The novel nozzle and path control methods were designed *to satisfy* the demands of continuous carbon fiber printing.

译文:为满足连续碳纤维打印的要求,设计了新型喷嘴和路径控制方法。

值得注意的是,不定式的被动式形式在通用英语中用得不多,但在专业英语中却频繁出现。另外,动词不定式作主语时,在很多情况下可使用形式主语。如形式主语 it 是没有任何词义的,绝不能把它理解成"它"。例如:

[4] It is possible *to have* interatomic bonds that are partially ionic and partially covalent, and, in fact, very few compounds exhibit pure ionic or covalent bonding.

译文:有可能是部分离子和原子组成的共价键,而且,事实上,很少有纯粹的离子键或共价键化合物。

[5] It is often necessary to express the composition of an alloy in term of its constituent elements.

译文:通常根据合金的组成元素来表述合金的组成。

[6] It should not be too surprising to find that a charge that is moving in a magnetic field experiences a force.

译文:发现在磁场中运动的电荷受到力的作用不应该是多么令人吃惊的事情。

2)动名词结构

动名词是一种非谓语动词形式,它的一般形式为"(been) done","having (been) done)"。

[1] Some materials are known as elastic because they return to their original shape after *having been bent*.

译文:有些材料被称为弹性体,因为他们经弯曲后仍能恢复原状。

[2] Cracking the HAZ was observed only in samples that had been prestrained before testing.

译文:只在测试前预应力的样品中观察到焊接热影响区裂纹。

[3] *Yielding* in pure shear occurs when the largest shear $\sigma_1 = k$ and $\sigma_3 = -\sigma_1 = -k$, where k is the yield strength in shear.

译文:当最大切应力 $\sigma_1 = k$ 和 $\sigma_3 = -\sigma_1 = -k$ 时,材料发生屈服,其中 k 是剪切屈服强度。

[4] This design maintains equal pressure into the mixing chamber *by using* the incoming supply of one gas component to pilot both inlet gas source regulators.

译文:该设计通过使用一种气体成分的进入供应来引导两个入口气体源调节器,以此维持进入混合室的等压。

[5] Small welded assemblies can be thermally stress relieved by heating the steel to 1,150 ℉, *holding* it for a predetermined length of time (typically 1h/inch of thickness), and allowing it to

return to room temperature.

译文:小的焊接组件可以通过将钢加热到1,150 ℉,保持预定长度的时间(通常为1h/inch 的厚度),并允许其恢复到室温,来消除热应力。

［6］High-strength aluminum alloys have the tensile strength of medium strong steel alloys while providing significant weight advantages.

译文:高强度铝合金具有中强度合金钢的抗拉强度,同时具有显著的重量优势。

［7］By the proper application of heat, as in annealing, these work-hardened metals can be made to recrystallize, making them softer, more ductile and more amenable to further manufacturing processes.

译文:正确的热处理(如退火),使加工硬化后的金属产生再结晶,从而使金属变得较软、塑性更好、更易于切削加工。

3)分词结构

分词的形式和动名词的形式一样。

［1］The remaining discussion of mechanical behavior assumes isotropy and polycrystallinity because such is the character of most engineering materials.

译文:其他的关于假设各向同性和多晶性力学性能的讨论,因为这是大多数工程材料的特性。

［2］The upsetting forming process and law of metal flowing was analyzed for large forming flange axle with two adjacent stepped axles.

译文:分析了法兰及相邻台阶轴镦粗过程及金属流动规律,并给出了解决毛坯稳定性和正确成形台阶轴的措施。

［3］Of all metals silver is the best conductor, followed by copper and aluminum.

译文:在所有金属中,银是最好的导体,铜和铝次之。

分词复合结构的逻辑主语一般是名词或代词,名词前面还可以加上 with 或 without 表示肯定或否定。分词复合结构在句中只作状语。例如:

［4］It cools as it rises, with the result of *having cold metal* in the risers and hot metal at the gate.

译文:随着其上升而冷却,结果是冒口中金属已经冷却,而浇口处还是液态金属。

［5］Having led the way in the race understand the new materials, most of the scientists wonder whether the United States will be the first to bring its invention to market.

译文:在认识到这种新材料在竞争中处于领先地位之后,多数科学家尚不知道美国是否会首先把自己的发明成果投放市场(时间状语)。

36.3　专业英语的词汇特点

英国语言学家 D. A. Wilkins(威尔金斯)在 *Linguistics in Language Teaching*(《语言教学

215

理论》）（1972）一书中指出："Without grammar very little can be conveyed, without vocabulary nothing can be conveyed."（如果没有语音和语法，还可以传达一点点信息，但是没有词汇，那就不能传达任何信息）。专业英语词汇量大且具有专业特殊性，要获得信息就必须有较大的英语词汇量。但是拥有的词汇量再大，若不熟悉该行业的专业词汇，也并不意味着他可以毫无困难地阅读和使用专业英语。因此了解专业词汇的特点、构词法及各种缩略词的造词法，对于学习掌握专业英语词汇是十分必要的。

36.3.1　专业英语词汇构成特点

1）新词汇层出不穷

21 世纪以来，随着新材料、新设备、新工艺的涌现和发展，新的专业英语词语层出不穷，往往率先通过科技交流在全球传播和应用。例如前缀 nano- 的出现说明了材料科学的新发展，以"纳米"为代表的新词语已经家喻户晓，如 nanotechnology（纳米技术）、nanoscience（纳米科学）、nanomaterial（纳米材料）。2004 年英国曼彻斯特大学的两位科学家安德烈·盖姆（Andre Geim）和克斯特亚·诺沃消洛夫（Konstantin Novoselov）发现了一种新型纳米材料，简称石墨烯（graphene），随后全世界的科技工作者掀起了研究石墨烯的热潮。graphene 是一个新词，是石墨（graphite）和烯（alkene）的合成词，graphene oxide（单层氧化石墨烯），functionalized graphenes（功能化石墨烯），bilayer graphene（双层氧化石墨烯），这些专业术语逐渐被人们所熟知。

2）通用英语词汇专业化

专业英语中使用的大量通用英语词汇，往往属于半专业性词汇（semi technical terms）。半专业词汇是指在各个学科领域里都经常使用的意义不尽相同的词汇，它们在专业文献中起着重要的修辞语篇功能，常被用来表达作者的意图和观点。半专业性词汇其实都是通用英语词汇，不过将其意义进行了扩展、延伸。像 eye, carrier, force, system, power, transmission, feed, run, work, energy, power 等既是通用英语词汇，又是半专业词汇。掌握了它们在通用英语中的含义，就容易了解它们在专业英语中的含义。例如"cast"在基础英语中，其基本词义是"演员"，专业英语中可以根据不同学科领域分别表示"铸造、铸件、手臂、腿"等。再如以下词汇在不同专业中词义有差别：element 元素（化学），天气（气候），电热丝（电工），自然环境（环境），seal 海豹（生物），密封垫（机械），封蜡模（印刷）。使用这些词汇时，必须根据不同场合去判断其词义，才不致引起误解。如"It may be assumed that a **strong** material is also **hard**."这样的句子，可以根据该学科领域译为："一般来说，材料**强度**高，其**硬度**也大"。

下面以"power"为例，不同的专业领域，有着不同的意义。

［1］In any case work does not include time, but **power** does.

译文：总之，功不包括时间，而**功率**包括时间。（物理）

［2］They've switched off the **power.**

译文：他们关掉了**电源**。（电工）

［3］Any number to the **power** of nought is equal to one.

译文：任何数的 0 **次方**等于 1。（数学）

[4] The planes are **powered** by Rolls Royce engines.

译文:这些飞机由劳斯莱斯公司制造的发动机**提供动力**。(机械)

3)专业英语词汇词义单一化

这是指各个学科或专业中应用的专业词汇或术语,其意义狭窄、单一,专业性很强,一般只使用在各自的专业范围内。这类词汇一般字母较多,通用英语中用得不多,字母越多词义越狭窄,出现的频率低。例如:

oxidation	氧化	semiconductor	半导体
hypereutectic	过共晶的	oscilloscope	示波器
electroluminescence	场致发光	decarburization	脱碳

36.3.2 专业英语中词汇的转换功能

专业英语中的有些词,看似都是日常用词,但用到专业英语方面经过转义、合成或派生,日常用语就被赋予了全新的含义。例如光学词汇 scatter(散射)其实就是常用词汇 scatter(散布、分散)的转义;radioactivity(放射性)就是 radio 和 activity 二词的合成;semi-conductor(半导体)就是由 conduct(引导)转义后再加后缀-or 和前缀 semi-派生而来。有些专业词汇对于只精通通用英语的人来说,可能是完全陌生的,如 pouring shot(注入槽), body core(砂芯主体), drag(后拖量), yield limit(屈服点)等这样一些专业词汇。而这些专业词汇,对于材料成型专业的学生来说,则是必须记住的。事实上,专业英语中的专业词汇是使专业英语区别于通用英语的最主要的内容,掌握专业英语专业词汇的构词规律和大量的专业词汇对提高阅读速度起很大作用。

1)**转化法**(conversion)

由一个词类转化为另一个词类,如名词转换为动词,转化后词形不变,只是词类转变。例如:

[1] It's possible *to weld* stainless steel to ordinary steel.

译文:不锈钢是有可能和普通钢**焊接**在一起的。

[2] By the analysis and measures above, it improves the test quality of the process piping butt *weld* effectively.

译文:通过以上分析和采取的措施,能有效提高工艺管道对接**焊缝**超声波检测的质量。

[3] The microstructure of isothermally-quenched ductile iron (ADI) is a mixture structure consisting of *austenite* plus acicular ferrite.

译文:等温淬火球铁的显微组织由**奥氏体**加针状铁素体的混合组织组成。

[4] The results showed that the heating rate plays a key role in affecting *austenite*-grain growth.

译文:结果表明,加热速率对**奥氏体化**晶粒长大有显著影响。

2)**合成法**(compound)

由两个或更多的词直接合成一个新词。在专业英语中这种构词法数量非常多,合成的类型有:名词+名词,形容词+名词,动词+副词,名词+动词,介词+名词,形容词+动词等。

例如：

workshop	车间	as-cast	毛坯铸件
oxy-acetylene	氧乙炔的	self-hardening	自硬性
shadowgraph	光摄影	superplasticity	超塑性
overflow	溢流	forgeability	可锻性
closed-die	合模	shot-blasting	喷丸清理
finish-cutting	精加工	form-turn	成形车削
grain-oriented	晶粒取向	self-diffusion	自扩散
close-grained	细颗粒的	horsepower	马力

3）派生法（derivation）

派生法又称词缀法，即在一个单词的前面或后面加词缀来构建新词的方法。这是英语构词的一种主要手段，对专业英语词汇的构成很有帮助。前缀和后缀的使用是创造派生词的常用方法。

（1）前缀

①常见的否定前缀。在英语中，有一些表示否定意义的前缀，在表示否定意义的时候，它们的用法基本相同，可译成汉语"非""无""不（是）""未"等。常见的否定前缀有：un-、non-、in-、dis-、de-、under-、anti-、counter-、mis-等。

表36.2 常见的否定前缀

前 缀	意 义	例 词
un-	不；未（=not）	unnealled 未退火的 undilluted 未稀释的 uncrystallized 未结晶的 unalloyed 没有杂物的
non-	非；不	nonconductor 绝缘体 nondestructive 不破坏的 noncontinuous 不连续的 nondeformable 不变形的
in- il- im-、ir-	不；非；无	inactive 不活泼的　　inaccurate 不精确的 imperfect 不完美的　　imbalance 不平衡的 irregular 不规则的
dis-	否定；相反；分离	dislocation 位错　dispersal 扩散、弥散 distortion 畸变
de-	除去	deoxidation 脱氧 degradation 退化；降解；分解
under-	不足	underquenching 淬火不足 undercure 欠硫化

前　缀	意　义	例　词
anti-	反对;抗	antimagnetic 抗磁的 anticorodal 耐蚀的
counter-	反对;逆	countershaft 副轴、中间轴
mis-	错误	misalignment 错位　mismatch 偏模

②表示空间位置,方向关系的前缀,见表36.3。

表36.3　表示空间位置和方向关系的前缀

前　缀	意　义	例　词
by-	附近,邻近,边侧	by-product 副产品
circum- circu-	周围,环绕,回转	circumscribe 外切 circumference 圆周
en-	置;使	envelope 包络线
ex-,ec-,es	外部,外	excite 激发　extrusion 挤压
up-	向上,向上面	upset 顶锻、镦粗
inter-,intel-	在……间,相互	interalloy 中间合金　intercrystalline 晶粒间的
intro-	向内,在内,内侧	intronuclear 原子核内的
med-,mid-	中,中间	midline 中线
out-	超过;向外	outline 轮廓线　outgate 溢流冒口
over-	超过;过分	overaging 过时效
per-	高	permanganate 高锰酸钾
post-	在后;补充	post-annealing 焊后退火
pre-;pro-	在前;向前	preformation 预成形　proeutectoid 先共析体
sub-,	在下面,下	sublattice 亚点阵、亚晶格
sur-,super-	在上;超	surfusion 过冷　superburning 过烧
trans-	横过;转移	transubstantiation 变质　transparent 透明
ultra-	超;极端	ultralumin 超硬铝
hyper-	超越	hyperdrostatics 流体静力学
hypo-	低;次	hypoeutectoid 亚共析的
peri-	周围;近;环境	peritectic 包晶(体)的

③表示数量关系的前缀,见表 36.4。

表 36.4　表示数量关系的前缀

前　缀	意　义	例　词
bi-	双;二	bimetallic 双金属的
centi-	百	centigrade 百分度的
co-	一起;共;和	cohesion 内聚力
deca-	十	decagon 十边形
deci-	十分之一	decimal 十进的;小数的;小数
di-	二	dioxide 二氧化物
equi-	同等	equilateral 等边的
half-	半	half-cell 半电池
hcto-	百	hectowatt 百瓦
hemi-	半	hemicrystalline 半结晶的
hetero-	异;杂	heteromorhous 多晶的
hex(a)-	六	hexagon 六角形　hexagonal 六方
homo-	同	homophase 同相
kilo-	千	kilowatt 千瓦
meg(a)-	大;兆;百万	megavolt 兆伏
micro	微;百万分之一	microhardness 显微硬度
muti-	多	mutiphase 多相的
oct(a,o)-	八	octagon 八角形
pent(a)-	五	pentane 戊烷
poly-	多	polyethylene 聚乙烯
quadr-	四	quadrate 正方形的
semi-	半	semiconductor 半导体
sex-	六	sexadecimal 十六进制的
sept-	七	septangle 七边形
tri-	三次;三倍;三	triode 三极管　triclinic 三斜晶系的
uni-	单;一	uniaxial 单轴的　uniphase 单相
re-	再次;重复	remelt 重熔　recrystallization 再结晶

④其他特定含义的前缀

还有一些用得比较少的前缀,它们也有特定的意义,见表36.5。

表 36.5　其他特定含义的前缀

前　　缀	意　　义	例　　词
ab-	脱离	ablation 磨削、烧蚀　abrasion 磨损
auto-	自己;自动	automobile 汽车
astro-	天文学的	astronaut 宇航员
electro-	电的	electrode 焊条　electrolyte 电解液
hydro-	水	hydrocarbon 碳氢化合物
photo-	光	photoeffect 光电效应　photocell 光电池
pyro-	火	pyrolysis 高温分解
radio-	放射;辐射	radioelement 放射性元素
tele-	远距离;电视	teleautomatics 自动遥控装置
thermo-	热	thermoplastic 热塑性
sph-	球	spheroidizing 球化处理　spherical 球形

(2)后缀

后缀是指在一个词的尾部加上一个词缀构成新的词,见表36.6。

表 36.6　常见后缀

后　　缀	意　　义	例　　词
-able	……的	exchangeable 可互换的
-al	有……属性的	artificial 人工的
-en	使……	soften 软化　harden 硬化
-er, or	……的人	fitter 装配工　compressor 压缩机
-graph	写、画、记录结果	monograph 专题;论文
-ilc	……的	metallic 金属的　chemical 化学的
-ics	……学	mechanics 机械学
-ify	使……	intensify 强化、加剧
-ism	……主义或学说	electromagnetism 电磁学
-ist	从事……者	physicist 物理学家
-ive	容易……的	conductive 传导的
-less	无	stainless 不锈钢
-logy	……学	technology 工艺学、技术
-ment	具体事物,工具	attachment 附件

续表

后　缀	意　义	例　词
-meter	……计、仪	voltmeter 电压表　micrometer 千分表
-ous	有……的	ferrous 铁的
-proof	防……的	acid-proof 耐酸的
-scope	……的镜	spectroscope 分光镜
-th	……度	length 长度
-ty,-ity	……性质、状态	density 密度

36.3.3　专业英语中的词语缩写(abbreviation)

把单词的音节按一定形式加以省略或简化而产生的词统称为缩略词,这种构词方法称为缩略法。专业英语中缩略词种类繁多,主要有四种类型:截短词(clipped word)、首字母缩略词(initialism)、首字母拼音词(acronym)、拼缀词(blend)。

1)截短词

截除原词的某一或某些音节所得的缩略词,称为截短词(clipped word),例如:

fluidics←	fluidonics	射流学	cub←	cubic	立方的
lab←	laboratory	实验室	cy←	cycle	循环;周期
di(a)←	diameter	直径	des←	design	设计
dyn←	dynamic	动力学的	elas←	elasticity	弹性
eq←	equation	方程	fig←	figure	图
freq←	frequency	频率	fun←	function	函数
grad←	gradient	梯度	ident←	identify	识别

2)首字母缩略词

利用词的第一个字母代表一个词组的缩略词,就称为首字母缩略词(initialism),缩略词多按字母读音,这种构词法在专业英语中颇为常见。

FCC—Face-Centered Cubic Lattice 面心立方晶格

M_f—Martensite finishing point 马氏体转变终了温度

CCT—continuous cooling transformation(CCT) curve 连续冷却转变曲线

LF—Ladle Furnace 炉外精炼

VOD—Vacuum Oxygen Decarburization 真空吹氧脱碳法

WPS—Welding Procedure Specification 焊接工艺规程

PVD—Physical Vapor Deposition 物理气相沉积

CE—Carbon Equivalent 碳当量

ASME—American Society of Mechanical Engineers 美国机械工程师学会

AISI—American Iron and Steel Institute 美国钢铁学会

AMP—Advanced Manufacturing Partnership 先进制造伙伴计划

MGI—Materials Genome Initiative 材料基因组计划

CMES—Chinese Mechanical Engineering Society 中国机械工程学会

3）首字母拼音词

首字母拼音词（acronym）与首字母缩略词类似，读音可以连读，例如：

SARS—Severe Acute Respiratory Syndrome 非典型肺炎

ROM—Read-Only Memory 只读存储器

DOS—Disk Operating System 磁盘操作系统

RADAR—radio detection and ranging 雷达

4）拼缀词

近年来，随着科学技术日新月异的发展，英语中表达新生事物的更多科技新词汇应运而生，这种人造的拼缀词（blend）在现代专业英语中日益增多。拼缀词一般把两个词各取一部分合在一起，构成一个新词，常见的有：

liar＝light＋radar 激光雷达

biotech＝biology＋technology 生物技术

smaze＝smog＋haze 烟霾

mocamp＝motor＋camp 汽车宿营地

hi-tech＝high＋technology 高技术

mechatronics＝mechanics＋electronics 机械电子学、机电一体化

econobox＝economy＋box 微型经济轿车

alnico＝alloy of aluminum, nickel and cobalt 铝镍钴合金

有些拼缀词，特别是新技术方面，很有生命力，为公众所接受并收入词典。掌握拼缀词及拼缀法对提高科技文献阅读和理解是大有益处的。

36.4 专业英语的语篇特点

什么叫作语篇？北京大学教授胡壮麟（1994：1-2）在谈及语篇时说："它可以是一个词……，一个短语或词组……，一个小句……，一副对联、一首小诗、一篇散文、一则日记、一部小说……，它也可以是一句口号、一支歌曲、一次对话、一场口角，一场长达两三个小时的演讲……非一言两语能够了事。因此多数场合下要动用好多句子，于是更多的人把语篇看成是大于句子的单位，尽管这样的理解不很精确。"本小节所定义的"语篇"是大于句子单位的语言片段，在阅读专业英文文献时，经验不足的阅读者往往将句子集中在句子层面上，每个专业术语，每个句子都理解得很正确，但对整篇文献的整体把握不足；经验丰富的研究者则会细心体会语篇的意义，对原文进行总体性思考，在语篇框架下对词汇和句子进行入情入理的斟酌，以体会作者新发明、新发现所使用的实验方法及实验结论等关键信息。

语篇是一个语言概念，它具有自己的特点。语篇不是一连串句子的简单组合或拼凑组

合,而是一个从头到尾贯穿着内在的逻辑结构的整体,有表达整体意义的特点。

36. 4. 1 衔接性

在语言形式上,语篇内各句、段之间存在着衔接性,英国当代语言学家 M. A. K. Halliday 把衔接分为语法衔接(grammatical cohesion)和词汇衔接(lexical cohesion)。语法衔接有四种:连接(conjunction)、替代(substitution)、省略(ellipsis)、照应(reference);词汇衔接也有四种:重复(repetition)、同义/反义(synonym/antonym)、上下义/局部——整体关系(hyponymy /metonymy)和搭配(collocation)。

1)照应

照应(reference)指用 it, they, this, these, that, those, here, now, then, there、the、same, identical, equal, similarly, likewise, so, as, such, so, more, less, equally, better, more 等形式来表示的语义关系,是语篇中某一种语言成分和另一种语言成分之间在指称意义上的相互解释的关系。照应可以使发话者运用间断的指代形式来表达上下文中已经或即将提到的内容,使语篇在修辞上有言简意赅的效果。更重要的是,它可以使语篇在结构上更加紧凑,使之成为前后衔接的整体。例如:

[1] The helium nucleus contains two other particles besides the two protons. Because they are neutral electrically, they are called "neutrons"。

分析:后一句中两个"they"均指"the two protons",形成人称照应。明确两个"they"所指后,这里可将其理解为:除了两个质子之外,氦原子核还包含其他两个粒子。这两个粒子由于是中性的,所以被称为"中子"。

[2] The welding industry has developed special high-deposition-rated core electrode wires. These wires contain little or no fluxing agent in the core. They have extremely high deposition efficiencies of 95% and greater, which are on the order of those of solid electrode wires.

分析:后两句中的"those"和"They"都指代"high-deposition-rated core electrode wires"。

[3] Homogenization of castings by normalizing may be done in order to break up or refine the dendritic structure and facilitate a more even response to subsequent hardening. Similarly, for wrought products, normalization can obliterate banded grain structure due to hot rolling, as well as large grain size or mixed large and small grain due to forging practice.

分析:前一句陈述了正火在铸件生产中的作用,为了上下句子的衔接,后一句使用副词"similarly"描述了正火在锻件生产中的意义。后一句可这样理解:同样,在锻造生产中采用正火工艺可以消除热轧中的带状组织,也可以消除锻造生产中的晶粒大小不均匀的现象。

[4] A number of gases will supply carbon to steel under proper conditions. One of the most commonly used combination is an endogas carrier gas enriched with about 10% natural gas as the carburizing agent. Another less commonly used mixture consists of about 80% nitrogen mixed with various proportion of natural gas and CO_2 as the carburizing material. These mixtures react directly with austenite to put carbon into solution and require only that a supply of the mixture be kept in contact with the steel over a period of time.

分析:该文段共四个句子,第一句提出许多气体将在适当的条件下向钢供应碳,第二句和第三句分别使用 most 和 less 举例,第四句中的 these 指代前面提到的两种混合物,说明这些气体发生渗碳的机理。从整个文段行文上看,层次分明,结构紧凑,前后相互衔接成整体。

　　2) 省略

省略(ellipsis)是把语篇中的某个成分略去不提,这一部分往往都是已知信息。它是为了避免重复,突出新信息,并使语篇上下紧凑的一种语法手段。例如:

　　[1] Cold-rolled steel is made from hot-rolled steel, that is, steel that has been rolled while hot, is then rolled again when cold.

分析:冷轧钢是由热轧钢轧制的。也就是说,在炽热状态下轧过的钢,冷却时再一次进行轧制。该句中 while 和 when 后均省略了 it is, it 代替 steel。

在连词 if, when, whenever, where, wherever 与形容词 possible, necessary 组成的结构中,通常认为连词后面省略了 it is。这里的 it 不是指某个词,而是指整个句子表达的内容。例如:

　　[2] Carbon steels are comparatively cheap and have many good mechanical properties, and we use them for general engineering purpose whenever possible.

分析:碳钢非常便宜而且具有很多优良的机械性能,因此,每当有可能,我们就把碳钢用于一般工程项目。连词 whenever 后面省略了 it is。

　　3) 连接

连接(conjunction)是通过连接成分体现语篇中各种逻辑关系的手段。连接成分往往是一些过渡性的词语,表示时间、因果、条件、转折、递进等逻辑上的联系。连接关系主要由连词、连接副词和介词短语等来体现,常用的有 and, but, because, anyway, however, following, in that case, by that time 等词。

　　[1] The result is that *equiaxed grains* (等轴晶) in a pure metal always adopt a dendritic morphology. Because segregation is absent, the dendritic form will not be detectable in a cast. (因果关系)

　　[2] *As before*, consider the behavior of a perturbation arising during directional solidification. (递进关系) Such a protuberances at the solid/liquid interface will increase the local temperature gradient in the melt(熔体). *In the case of* a pure melt(Fig. 3.2), this led to the disappearance of the perturbation. (条件关系) However, in an alloy melt(Fig. 3.4), the local concentration gradient will also become steeper and consequently the local gradient of the liquidus temperature will increase. (转折关系) Hence, the region of constitutional supercooling will tend to be preserved. (因果关系)

　　4) 替代

替代(substitution)是用替代词取代上下文出现的词语。替代既可以避免重复,也可以连接上下文。替代有名词替代、动词替代和小句替代三种。当某个同类事物出现时由相应的语法项目(如 one, ones, the same 等)来替代的现象为名词替代;当某个同类动作、行为出现时由相应的语法项目(如 do)替代的现象是动词替代;小句替代指当一个同类小句出现时由相应的语法项目(如 so)来替代的现象。例如：

[1] Some soft varieties of limestone contain from 6% to 14% bitumen, and a few crystalline ones contain 2% to 20%.

分析：后半句中的 ones 指代 limestones。

5)词汇衔接

词汇衔接(lexical cohesion)主要通过重复、同义/反义、上下义/局部——整体关系和搭配等词汇手段来体现语篇的语义连贯。词汇的重复指语篇中单词和词组的重复。词汇重复可以使语篇连贯，给读者留下深刻印象，从而达到突出主题的目的。"同义性"，也指各种形式的"反义性"。同义性指具有同样意义的不同词项之间的接应关系；反义性是同义性的极端，表示不同程度性质的词语，体现对立的层次性。上下义词是相对而言的。上义词的含义概括而抽象，下义词则表示比较具体的含义，一个上义词往往包括若干个"共同下义词"。词汇同现即词汇搭配也具有连句成篇的作用，它可以出现在同一个小句内，也可以出现在不同的小句内。

[1] However, other methods exploit natural forms of gene transfer, such as the ability of Agrobacterium to transfer genetic material to plants, and the ability of lentiviruses to transfer genes to animal cell.

分析：在这句话中，transfer 一词重复了三次，the ability of 词组重复了两次，gene 一词也出现了两次，通过词汇的简单重复所形成的语篇纽带作用，使句子前后呼应，强调和突出了主题，并且有效地加深了读者对"基因转换"这一概念的认识和对语篇内容的理解。

[2] Typically, genetically modified foods(转基因食品) are transgenic plant products：soybean, corn, canola, and cotton seed oil.

分析：在这个概念中 soybean, corn, canola 与 cotton seed oil 在 genetically modified foods 这个共同概念支配下结合在一起，形成一个语义场，genetically modified foods 与它们构成一种上下义关系，彼此呼应前后衔接。而且，通过 genetically modified foods 这一概念的上下义词复现，使我们能更清楚地了解转基因的概念。

36.4.2　连贯性

如上所述，衔接是语篇中不同语句成分之间的语义联系，是语篇的有形网络，而连贯是语篇的无形网络，它是指一段话语或某个语篇的不同部分在意义上的联系。Halliday & Hasan 把衔接与语篇连贯联系起来，认为如果某个语篇具有连贯性，这种连贯性势必通过语言本身得到反映，反映的一个重要途径便是语言的衔接，衔接同层次相互联系的语义关系，另外连贯也可能通过语境和逻辑等手段实现。它是情景语境与语言形式相互作用的结果。更具体地讲，它由情景语境决定，由形式(词汇、语法、语音)来体现。从情景语境的角度讲，如果语篇在情景语境中行使适当的功能，它就是连贯的。它意味着语篇与情景语境的关系已经建立起来。从语篇本身来讲，它表示语篇的语义连接完好，语篇的各个部分在整个语篇中起作用，形成一个语义整体。这时，我们必须把语篇发生的具体环境，从语篇发生的语言环境中通过逻辑推理来推断其在语义上的连贯。

36.4.3　意向性

意向性起源于哲学,是主体的心理状态借以指向或涉及其自身之外的客体和事态的那种特征。约翰·塞尔首次将言语行为与意向性结合起来,提出"每一个有意义的语句借助意义可以来施行一种特定的言语行为,而因为每一种可能的言语行为原则上可以在一个或若干个语句中得到表述。因此,语句意义的研究和言语行为的研究不是两种不相关的研究,而是一种从不同角度进行的研究。"每个语篇都有一定的目的性,它总是想把其意图告诉读者,这就是意向性。一般而言,科技语篇的意向性决定于科技语篇的类型。

第一类是论证型——对基础性科学命题的论述与证明,或对提出的新的设想原理、模型、材料、工艺等进行理论分析,使其完善、补充或修正。

第二类是科技报告型——科技报告是描述一项科学技术研究的结果或进展,或一项技术研究试验和评价的结果,或论述某项科学技术问题的现状和发展的文件。

第三类是发现、发明型——记述被发现事物或事件的背景、现象、本质、特性及其运动变化规律和人类使用这种发现前景的文章。

第四类是设计、计算型——为解决某些工程问题、技术问题和管理问题而进行的计算机程序设计,某些系统、工程方案、产品的计算机辅助设计、优化设计以及某些过程的计算机模拟,某些产品或材料的设计或调制和配制等。

第五类是综述型——一种比较特殊的科技论文。它的写法通常有两类:一类以汇集文献资料为主,辅以注释,客观而少评述。另一类则着重评述,通过回顾、观察和展望,提出合乎逻辑的、具有启迪性的看法和建议。例如:

[1] The development and research of titanium cast alloy and its casting technology, especially its application in aeronautical industry in China are presented. The technology of molding, melting and casting of titanium alloy, casting quality control are introduced. The existing problems and development trend in titanium alloy casting technology are also discussed.

分析:这是一篇 *Development and application of titanium alloy casting technology in China* 专业英语的摘要,属于上述第五类综述型科技论文。对钛合金铸造技术的现状、质量控制、发展以及存在的问题进行总结分析,向专业读者传达有效的信息。

37

英汉两种语言的对比

英语和汉语分属于印欧语系和汉藏语系,由于形成和发展历程各异,英汉语言各具特点。英语国家沿袭了古代希腊非常严格和规范的语词系统。古代希腊人认为,语词系统与思维系统是一致的,要表达一个清晰合理的思想就离不开清晰合理的词形和句法。与之相反,中国人重直觉,强调"悟性",只要能够达意,词的形式无关紧要,词语之间的关系经常在不言中,语法意义和逻辑关系常隐藏在字里行间。从语言学的角度来说,英汉语言之间最重要的区别特征莫过于形合(hypotaxis)和意合(parataxis)之分(尤金·A. 奈达,1982:16)。连淑能《英汉对比研究》(增订本)指出"形合是句中的词语或分句之间用语言形式手段连接起来,表达语法意义和逻辑关系。意合是词语或分句之间不用语言形式手段连接,句中的语法意义和逻辑关系通过词语或分句的含义表达。"

不同语言的对比分析,不仅有利于阅读和翻译,也有助于语言交际。通过对比分析,我们可以进一步认识外语和母语的特性,在使用时,能够有意识地注意不同语言的不同表现方法,以顺应这些差异,防止表达错误,避免运用失当。

37.1　英汉词汇对比

《现代汉语词典(第6版)》收词6.9万,常用的汉字3,500个左右;《牛津高阶英汉双解词典(第8版)》收纳了184 500项单词、短语和释义。汉语新词语虽然也层出不穷,但极少有新字产生。偶有新字产生,如《现代汉语词典(第7版)》较第6版增新词语仅400多条,汉字新字常用构词方法往往是偏旁部首加谐音字根一并便成了。这种构词法不像英语那样不断衍生,靠的是固有的汉字的灵活组构,因此汉字总量总能保持在一定数量之内。随着时代的发展,常用汉字不但没有增长,反而有所减少——一些老字不断被赋予新意,而更多的古字则逐渐被时代所淘汰。而英语使用各种构词方法,新词汇层出不穷,其总词汇量早已超过百万。中国人如果掌握了3 000~4 000个汉字,就可以读懂许多报纸杂志了,而英文如果只知道10 000词还远远不够。汉语词义较概括,而英语中的名词则定名具体,一词一物,很少有汉语中概念性的上坐标词语。词汇量对应方面,英译汉时,每1 000个英文词对译成1 720~1 790汉字;汉译英时,每1 000个英文词对应1 330~1 410汉字。从词汇语义学来看,英语和汉语的主要差异有三种情况:词的意义、词的搭配能力和词序(语序)方面。

37.1.1　词语的意义

英语词语有如下四个特点:词是不可分割的单位;词可以由一个或几个词素组成;词通常出现在短语结构中;词应该属于某个词类。一个词在语言中的使用就是它的意义。词语的意义是作者或者说话人试图传达并期望"被理解"的东西,常常与他们的(某种)意图相关。也就是说理解一个词的具体含义,必须根据具体的语境去分析。英语词汇意义在汉语中的对应程度,大致可归纳为三种情况。

1)词义完全对等

英语词义与汉语词义在任何上下文中都完全对等。这主要是一些已有通用译名的专用名词、术语等。例如:

manganese	锰	voltage	电压
CNC	数控机床	Rockwell hardness	洛氏硬度
Acrylic alloy	亚克力合金	carbide	碳化物
quenching	淬火	velocity	速度

2)部分对等

英语中有些词与汉语中有些词在词义上只有部分对应。它们在词义上概括的范围有广狭之分。如英语中的"die"一词,在汉语中可泛指各类模具,包括注塑模、冲压模、压铸模、挤压模、吹塑模等等。而"mould"仅指"铸模"。同样,汉语中的"工艺"一词,在英语中有"technology","procedure","technique","process","craft","workmanship"等,在不同的行业和语境中都可以理解为"工艺"。

3)词义空缺

语言之间的词义空缺是一种普遍存在的现象。英汉语中各自有不少属于各自所特有的词语。语言转换时,却找不到对应词汇的内涵。比如我们比较熟悉的例子 engine 引擎、motor 马达、Moore 摩尔、nanometer 纳米、troostite 托氏体等。

37.1.2　词语的搭配

英汉两种语言在词义对应、用词方式、思维模式等方面均存在差异,而这些差异也影响了英汉词语的搭配。英语词语搭配通常避免"同义反复";而汉语词语搭配却往往使用"同义反复"。例如:round *in shape* 圆形的、small *in size* 小型的、few *in number* 数量很少的、many *in number* 数量很多的、red *in color* 红色的、bitter-*tasting* 苦味的、*of an indefinite nature* 性质不明的、usual *habits* 通常的习惯、first *beginning*、to mix *together*、future *prospects* 未来的前景(注意:斜体加粗部分是英语使用中要避免的)。

英语中一词与不同的词搭配使用时,汉语的含义不同。如英语中"heat"一词作为名词"热、高温、温度"等,与下列不同的词搭配时,具有不同的含义,例如:

heat capacity	比热容	heat shortness	热脆性
heat source	热源	heat conductivity	导热性
heat treatment	热处理	heat number	炉号
heat retaining	保温	heat resistance	耐热性
heat shield	遮热板	heat checking	热裂
heat sink	散热体	heat plate	加热板
heat value	热值	heat transfer	热传递
heat shock	耐震	heat tinting	热蚀法

37.1.3　词序不同

语序是指"词、短语、句子、句群、段落和篇章中语素、词和语句的排列次序"。语序既是重要的语法手段,也是重要的修辞手段。不同的民族,往往因其思维习惯不同,而对同一客观事实有着不同的语言传达顺序。英语民族的思维反映现实的顺序主要是:主体→行为→行为客体→行为标志。这一思维习惯所引起的语言传达模式是:主语+谓语+宾语+状语,以及较长的定语必须后置等。汉语的思维方式则是:主体→行为标志→行为→行为客体。这一思维习惯所引起的语言传达模式是:主语+状语+谓语+宾语,以及宾语必须前置等。英语与汉语在民族思维习惯及语言形态上是有同有异,在基本语序上是大同小异。"同"的是主语、谓语和宾语的位置,"异"的是状语和定语的位置。例如:

〔1〕Without interchangeable manufacturing, modern industry could not exist, and without effective size control by the engineer, interchangeable manufacturing could not be achieved.

译文:没有可互换性制造,现代工业就不可能存在;没有工程师对零件尺寸的有效控制,可互换性制造也就不可能实现。

分析:该句中由介词 without 引出两个假设条件句,分别在句中作状语,英语中,状语的位置比较灵活,可以位于句首、句末或句中;汉语中汉语的状语往往在动词前。

〔2〕Parts made by transfer molding have greater strengths, more uniform densities, closer dimensional tolerances, and the parting lines requires less cleaning as compared with parts made by compression molding.

译文:与压缩模相比,传递模生产出来的塑件不仅具有较高的强度、比较均匀的密度及较为精确的尺寸公差,而且分型处需要清理的飞边、毛刺也少。

分析:made by transfer molding 和 made by compression molding 为分词短语作后置定语修饰定语 parts 一词。可以充当前置定语的有形容词、代词、数词、名词或名词所有格、动词的-ing 形式等。可以充当后置定语的有形容词、副词、介词短语、不定式、动词的-ing 形式、动词的-ed 形式。

37.2　英汉句子对比

英语造句常用各种形式手段连接词、语、分句或从句，注重显性接应（overt cohesion），注重句子形式，注重结构完整，注重以形显义。英语句中的连接手段和形式（cohesive ties）不仅数量大，种类多，而且用得十分频繁，常用关系词、连接词、介词，如 that，what，when，where，why，how，and，however，as well as，while，since，until，yet，of，between，in，according to，along with，throughout，it，there 等。英语造句几乎离不开这些关系词、连接词、介词，而汉语少用甚至不用这类词。汉语造句注重隐性连贯（covert coherence），注重逻辑事理顺序，注重功能、意义，注重以神统形，因而比较简洁。请看下面的例句：

[1] Solidification is of such importance simply because one of its major practical applications, namely casting, is a very economic method of forming a component if the melting point of the metal is not too high.

译文：凝固非常重要，因为其主要实际应用中的一个便是铸造，如果金属的熔点不太高，是一种非常经济的零件成形的方法。

分析：国内一些学者曾把英语句式形象地比喻为树式结构，指出，英语中，主干结构突出，即主、谓、宾结构突出，犹如一棵树的主干。英语在表达较复杂的思想时，往往开门见山，先把句中的主语和主要的动词这两根主枝竖起来，然后再运用各种关系把定语从句及其他短语往这两根主枝上挂钩。词句中 Solidification is of such importance…是主干部分，为什么重要？由 because 引导原因状语从句说明，英语中的原因分句前的从属连词一般是必须使用的，根据语言形式就能准确判断主从句的句间关系，因此英语是重意义也重形式，语言形式在很多情况下是必不可少的。of forming a component 短语在这作后置定语修饰前面的名词 method，if 引导条件状语从句。英语的长句比较多，往往从句里面含从句，短语里面含短语，如果没有弄清结构关系，对其语义的理解就会容易出错。

[2] When the ultimate strength of the material is reached, the fracture progresses, and if the clearance is correct and both edges are of equal sharpness, the fractures meet at the centre of the sheet as shown in Fig. 14-1(c).

译文：当材料达到了它的极限抗拉强度时，裂纹就会产生，如果间隙合适，(凹凸膜)刃口同样锋利，上下裂纹相遇在如图 14-1(c)所示的板料中央。

分析：先分析句子的结构、形式，才能确定句子的功能、意义。英语句子中，分句与分句之间一般用连词连接；而汉语句子中，分句之间有时用连词连接，有时却不用。在英汉两种语言中，连词不仅具有连接功能，而且有语义功能，能够表达词、分句、句子以及段落等语言单位之间的各种语义关系。此句是由连接词 and 连接的并列句，前句是由 when 引导的时间状语从句，后句是由 if 引导的条件状语从句。

[3] Precision grinding is the principle production method of cutting materials that are too hard to cut by other conventional tools or for producing surfaces on parts to tolerance or finish requirements more exacting than can be achieved by other manufacturing methods.

译文:精密磨削是加工传统切削工具难以加工的硬质材料的主要方法,这种加工方式所能达到的公差或表面粗糙度要高于其他加工方法。

分析:英语句子主要以主谓结构为纲,除少数省略等特殊情况外,在句子中主语不可缺,不管起不起作用总会有一个主语起统领全局的作用。在句子各种成分中,主语使用频率最高。主语是谓语描述的对象,位置在句首,在句中占据重要的位置。所有的主语必须是名词性的,这一点与汉语有很大的不同。其他词类要做主语,必须通过构词变化,变成名词,如:grind 磨削是动词,要变成 grinding。英汉定语的位置不完全相同。英语单词作定语时,有时定语放在名词之前,也有一些作后置定语,短语和从句作定语也要放在所修饰词的后面。而汉语的定语位置一般都放在所修饰词语的前面,当英语定语修饰的是由 some, any, every 等不定代词构成的复合词时,定语要后置。句中 materials 是先行词,that 是关系代词,引导定语从句修饰 materials。

[4] Thus, instead of expending energy against the typically high flow stress of a solid metal during forging or similar processes, it is only necessary to contend with the essentially zero shear stress of a liquid.

译文:因此,不像锻造或类似过程中固体金属的高流动应力消耗能量,(铸造)只需考虑的是流体最低剪切应力。

分析:instead of 是个短语介词。instead of 的意思是"代替……""而不……", instead of 功能与连词十分相似。it 作形式主语,代替后面真正的主语动词不定式 to contend with,专业英语中常用 It+be+形容词或"a(n)+名词"+to do sth. 的结构。语言学家 M. A. K. Halliday 认为 it 在专业英语中,用相对客观的方式委婉地传达了作者的情感和态度,避免了使用"I think""I believe"这样的表达,既没有违背专业英语的基本特征,也巧妙地传达了作者的意思,使得专业英语语篇客观而不失严谨。

37.3 英汉思维对比

思维模式受不同文化、个人知识结构、社会中工作环境及习惯的影响,因此,思维模式通常具有深厚的民族文化渊源。不同文化的人对外界认知模式的差异,往往导致思维模式的差异。1971 年周恩来总理和美国国务卿基辛格在讨论起草"中美上海公报"时,基辛格曾经说过,东方人的思维习惯是"异"中求"同",西方人是在"同"中求"异"。这句话在一定程度上反映了东西方民族不同的思维方式和思维视角。

由于思维与语言是相辅相成的,语言语法结构方式可以说是该民族思维方式的语言表现。从总体上看,传统的中国文化思维方式具有较强的具象性,而西方文化思维方式则具有较强的抽象性。汉字上承陶符,下启金文,字符象形,其起源的形象是原始图画,经后世演化,图画改为线条即成为象形文字,凸显简单的物象。西方民族的文字逐渐形成了概括某一类物象的概念语符,不像汉语文字那样直观形象。这种思维方式的差异反映在语言表达方式上,汉语用词倾向于具体,常常以实的形式表达虚的概念,以具体的形象表达抽象的内容;英语用词则倾向于虚。

由于思维差异,英汉语中动词、名词和介词的使用频率不同。英语是名词占优势的语言,汉语是动词占优势的语言。英语大量使用抽象名词和介词,因而显得虚、静和抽象;汉语多用动词,因而显得实、动和具体。英语重形合,导致动词的使用受到限制,而名词、介词、形容词、副词的使用非常活跃,表现力特强。汉语重意合,动词的使用不受限制。因此在阅读和理解时,就要作相应的词类转换,使其抽象性弱化,以适应汉语的表达方式。例如:

[1] Despite all the improvements, rubber still has a number of limitations.

译文:尽管改进了很多,但合成橡胶仍有一些缺陷。(名词理解为动词)

分析:有些英语名词是由动词演变来的,具有动作的意味,improvement 是名词,表示改进的意思,它是由动词 improve 加名词后缀-ment 变来的,这种情况可以将名词理解为汉语动词,显得简洁、自然。

[2] If extremely low-cost power were ever to become available from large nuclear power plants, electrolytic hydrogen would become competitive.

译文:如果能够从大型核电站获得成本极低的电力,电解氢的竞争能力就会增强。(形容词理解为动词)

分析:英语中的形容词也可以根据不同情况转换为其他汉语词性,使得译文合乎表达,如将形容词转译为动词、名词、副词等。

[3] The cutting tool must be strong, hard, tough, and wear resistant.

译文:刀具必须有足够的强度、硬度、韧性和耐磨性。(形容词理解为名词)

分析:当形容词指代事物的性质且作表语或前置定语时,形容词转换成名词,以体现科技类文体的专业性、科学性。

[4] With its many characteristic properties, iron has come to be the leading material for general engineering structures.

译文:铁具有多种性能,所以它已成为一般工程结构中的主要材料。(介词理解为动词)

分析:介词是英语中最为常见的一种词性之一,它不仅数量多,而且使用范围和使用频率也几乎超过其他词性的词。相比之下,汉语的介词没有那么多,使用也没有那么广泛。所以,在阅读和理解过程中,介词往往具有较强的动词意味,通常情况下,都可以转换为动词词性。

[5] The fatigue life test is over.

译文:疲劳寿命试验结束了。(副词理解为动词)

分析:英语中一些副词on, off, over, up, in, out, behind, forward 等充当表语、状语或补语,表示一系列的动作,而汉语的动词使用频率非常高,因此在描述动作时,通常会有动词。在阅读和理解时,就出现许多副词变成动词的情况。

[6] All structural materials behave plastically above their elastic range.

译文:超过弹性极限时,一切结构材料都会显示出可塑性。

分析:汉语科技文献中为了体现科学性和专业性,常常使用"……性""……度""……期"之类的表达,而英语专业文献中许多动词的副词恰恰描述了事物的某种特质,可以理解为上述的几种措辞,此时将副词转换为名词。

38

专业英语翻译

38.1 专业英语翻译概述

38.1.1 专业英语翻译的定义

在了解什么是"科技英语翻译"之前,我们必须清楚"翻译"的内涵。英文"translate"来自拉丁语"trans+latus",意思是"运载"(carried across),字面意义指把一种语言所表达的思维内容用另一种语言表达出来的跨语言、跨文化的语言交际活动。

中国历史上的翻译活动,先后形成了宗教翻译、文学(艺术)翻译与科学翻译三个体系。宗教翻译主要指佛教、基督教、伊斯兰教等宗教内容的翻译,汉唐与明清曾先后经历过多次发展鼎盛时期。文学(艺术)属于社会科学的范畴,文学翻译的发生并不比科学翻译更早。但是,在中国近现代时期,文学翻译发展得很快,队伍壮大绵延不断,翻译大家群星闪烁,理论研究日臻完善,翻译成果举世瞩目。科学翻译不仅包括自然科学、技术工程范畴的翻译,还涵盖了哲学及除文学艺术之外所有社会科学范围的翻译。

就中国科技英语翻译而言,可追溯到明末清初,从徐光启与西方传教士利玛窦合作翻译当时最有名的根据德国数学家克拉维斯编注的欧几里得《几何原本》的前六卷开始。到了中国近代,随着西方列强的入侵,中国的知识分子及仁人志士看到了学习西方近代科学技术的重要性。清朝政府机构开展了大规模、系统的科学翻译工作,成立了专门翻译的机构,如京师同文馆(1862 年)、江南制造局(1865 年)等,同时也出现了一批优秀的翻译工作者,如李善兰(1811—1882)、华蘅芳、赵元益(1840—1912)、何燎然、徐寿及其儿子徐建寅等。在大量的翻译实践过程中,逐渐形成了一些翻译理论、方法以及技巧。他们的翻译工作为 20 世纪中国的科学及科学思想的迅速发展铺平了道路。第二次世界大战以后,全球经济迅猛发展,科学技术日新月异,国际贸易、金融保险、邮电通信、国际旅游、科技交流等全球范围内的各种交往空前频繁,英语成了国际交往中的主要通用语言,专门用途英语(ESP)应运而生。

科技英语(EST)是专门用途英语的重要分支。随着改革开放的深入,国际科技交流与合作、科研成果介绍、大型学术专著与工具书的编写、科技资料翻译,技术转让、产品宣传、商务谈判等掀起了中国翻译的第四次高潮,并以现代语言学作为研究的理论基础,出现了奈达语言学——符号学翻译理论、乔姆斯基的转换生成语法、韩礼德的系统功能语言学、纽马克的语义/交际翻译理论等。

38.1.2 专业英语翻译的标准

对于翻译的标准,中国近代翻译大家有着不同的见解。严复的“信、达、雅”,鲁迅的“信顺”,郭沫若的“翻译创作论”,林语堂的“翻译美学论”,朱光潜的“翻译艺术论”,茅盾的“意境论”,傅雷的“神似”,钱钟书的“化境”,以及焦菊隐的“整体论”等。其中严复是我国翻译史上明确提出翻译标准的人,他在《天演论译例言》中提出:“译事三难:信、达、雅。顾信矣,不达,虽译,犹不译也,则达尚焉。”在中国,尽管这一标准,一直是人们争议的对象,但这也是影响最深、流传最广的翻译标准。“信、达、雅”同样适用于专业英语的翻译。

信,就是忠实于原文,不歪曲、不臆造。科学技术对于忠实原文的要求是特别严格的,大到理论的阐述,小到数据的举证,都不能有丝毫的谬误和误差。这就要求科技翻译必须做到准确,要求确切无误地表达原文本的技术路线和结论,不能有任何模棱两可之处。

达,是用规范、通顺的语言来表达原文中完整的意思,读起来通顺畅达、可读易懂。“达”是保证“信”的基本条件,它要求词语的选择、组合和搭配要恰到好处。句子的语序要恰当排列,各句之间语义逻辑紧密衔接,句型结构流畅,能准确地传达原语的情态、时态、语态等。

雅,是继“信”“达”之后对好的翻译的更高要求,也是译出好的译文或美文的必要条件。在科技英语中,雅要求译文尽可能简练,没有冗词废字。换言之,译文应该在准确、通顺的基础上,力求做到简洁明快、精炼概要。

总之,忠实、通顺、简练已成为公认的专业英语翻译标准。专业英语的翻译,必须全面地考虑原文中上下文之间的各种因素,如因果关系及文化背景差异等,才能找出与原文内容等值的表达方式。在翻译的过程中,译者必须把“忠实”原作的内容放在第一位,同时还要通顺、易懂、符合规范。由于语言受时间、空间的制约,而且不同时代的读者的接受能力不同,译者必须注意不同时期语言的变化,考虑到读者对语言的接受程度。

38.1.3 翻译的过程

对于翻译的步骤,奈达将之分为四步:一是分析(analysis),从语法和语义两方面对原文信息进行分析;而是传译(transfer),将经过分析的信息从原语译成译语;三是重组(restructuring),把传译过来的信息重组成符合要求的译语;四是检验(testing),对比译文意义与原文意义是否对等。国内的学者通常把翻译的步骤分为理解、表达、校对三个环节。这种阶段的划分对专业英语的翻译更具有指导意义。

1)理解

理解是翻译活动的开始,对原文的正确分析与理解是翻译的基本条件,也是最为重要的一步。在这一阶段中,译者要注意从不同角度对原文作好分析工作,包括对文章的词、句、段进行分析,由表及里,由词汇语法结构至语义及语境含义,从而求得对句、段话语含义的准确理解。对于一些专业英语文献词、句、段的理解,还应结合文献中提供的图表、数据、引用文献等资料进行分析,否则任何疏漏都会导致译文失准,甚至有时会与原文大相径庭。

2)表达

翻译的最终目的,是为了让读者了解原作的意思。表达的好坏直接影响到译文的质量。英文和汉语是两种不同的语言,在词汇和语法结构上有很大的差异。译者往往是要摆脱原句的结构和束缚,用符合目的语的习惯句式来表达原句的意思。在这种情况下,表达是至关重要的。只有正确理解原文,用恰当的译语形式把原文的意思通顺地表述出来,才能做好翻译工作。

3)校对

校对阶段可大致分为三个步骤进行。

第一步,着重检测翻译内容,包括译文有无漏译和错译现象,以及译文在字词句等各个语言层面是否全部准确无误地传达了原文信息。

第二步,核查译文在效果上是否做到与原文风格尽量贴近。

第三步,着重于译文的润色和美化工作,使翻译的内容尽可能简练,符合读者的阅读习惯。

38.1.4 翻译工作中所具备的条件

专业英语翻译是一项非常艰苦复杂的过程。关于翻译工作者必须具备的条件,大翻译家郭沫若指出必须具备四个条件,即外语知识、科学专业知识、汉语"自操纵的能力"以及责任心。

1)深厚的语言功底

茅盾先生说过,"精通本国语文和被翻译的语文,是从事翻译工作的起码条件"。扎实的英语功底是准确深入"理解"原文本的前提,深厚的汉语基础则是成功自然"表达"的重要条件,对英汉语异同的熟悉程度则进一步决定了译文最终的质量。

2)丰富的专业知识

科技翻译的客体是科学技术文章,因此译者必须拥有一定的科技知识。这里所谓的"丰富的",就是说要拥有尽量广泛的基础科技知识,掌握尽量多的技术术语和专业词汇,熟悉所译专业的基本科学概念。例如:

[1] A car with a manual transmission is assembled.

译文:正在装配一辆有手动变速器的汽车。

[2] Television is the transmission and reception of images of moving objects by radio waves.

译文:电视是通过无线电波传播和接受活动物体的图像的。

[3] Its transmission density is not known.

译文:其透射密度不详。

[4] The cause for the transmission for generation to generation of that disease has not yet been found.

译文:那种疾病代代遗传的原因,至今仍未被查明。

分析:上述 4 个例句中,"transmission"在例 1 句中机械学的意思是"传动"或"变速",在例 2 句中无线电、工程学中的意思变为"传播",在例 3 句中物理学中词义为"投射",在例 4 句中医学中词义变为"遗传"。

3)一定的翻译技巧

翻译实践表明,掌握一定的翻译技巧对科技翻译工作者来说不可或缺。所谓翻译技巧,就是在弄清表达同一意义的外语和汉语异同的基础上,找出处理其不同处的典型手法和转换规律,具体地说,就是在处理外语词义、词序、句型和结构等时所采用的手段。必要的翻译技巧不仅有助于弥合中外两种语言在表达方面的差异,为自己的汉语语言优势找到用武之地,而且更重要的是可以打造出毫无翻译痕迹、类似于汉语创作般的理想译文。

[1] The maglevtrain accomplishes the function of support, guidance, acceleration and *braking* by using *non-contact*—electromagnetic instead of mechanical force.

译文:磁悬浮列车**不与铁轨接触**,因而靠电磁力,而不是靠机械力来实现支撑、导向、加速和**刹车**功能。

分析:在日常生活中,"braking"翻译为"刹车"。但是在科技文体中,"刹车"用在该语境不是很妥帖,如果改成"制动",则符合科技文体的行文风格。"no-contact"直接翻译成"不接触"跟"磁悬浮列车"名称脱节,如果译成"悬浮于",就显得直观、易懂。

4)严谨、认真的工作态度

专业英语中不仅长句、难句较多,而且还经常会出现数字、数据、公式、方程式、各式符号、标记、图表等,所有这些都必须准确无误地出现在译文中,不能因有任何疏漏和错误而贻误工作。因此,在翻译过程中必须养成严谨、认真的工作态度,切不可由于翻译上的疏误,给科研或生产造成损失。

38.2 专业英语词汇的翻译方法

对于初学翻译的人来说,由于没有掌握基本的翻译方法容易出现两方面的问题。一方面,译出来的东西比较死板。这主要表现在死搬词典释义(往往是英汉词典)和对原文句子结构的亦步亦趋上。另一方面,初学者过于放开手脚,有时几乎完全脱离了原文,胡译乱译。好的翻译应该是严而不死,活而不乱。要想学好翻译,就必须掌握翻译的基本方法和技巧。词义的选择是翻译过程中首先要解决的问题。英语词汇量很大,它吸收了大量的其他语言中的词汇(主要是拉丁语、希腊语、法语),因此具有一词多义、同义近义词多等特点;而汉语的词的含义范围相对比较窄。此外,英语词汇所表达的意义并非与汉语完全对等,会出现缺位现象,在英译汉过程中准确把握词义、适当运用引申、词类转换、增译、减译、省略等方法是保证译文准确和通顺的前提和基础。

38.2.1 词义的选择

专业英语中词汇具有通用英语词汇专业化的特点,称之为半专业词汇,这些词汇保持着一词多义、一词多类、灵活多变的特点。这部分词汇和作为整个语言基础的普通词汇在文献中占有极大的比例。翻译时,只有选择适当的词,才能把它们的确切含义表达出来。

1）多义词词义的选择

英语跟汉语相同,根据不同的搭配或上下文,一个词可以表示多种意义。初学翻译时,我们对词的涵盖范围、搭配能力不十分清楚时,需要勤查字典,弄清它们在具体场合下的具体含义,再下笔翻译,以免产生望文生义的现象。

下面以常用词 take 为例进行说明:

take 既可以作为动词,也可以作为名词,这里就其作为动词的用法,举一些例子。这时它的中心词义是"拿""取""带",但用在不同场合还表示其他很多意思,因此要采用不同的译法。例如:

［1］Underwater Welding Schools：*Take* the Dive into A Specialty Career.

译文:水下焊接学校:潜入专业生涯。

［2］Full advantage is *taken* of fiber flow lines.

译文:充分利用纤维流线。

［3］It is its coating that *takes* the wear.

译文:是它的表面涂层承受磨损。

take 后面接表示时间、人力、能量之类的名词时,译成"花费""需要""用"。

［4］It *takes* a body precisely as long to fall from a height *h* as it does rise that high.

译文:一物体从高 *h* 下落*所需*的时间与它上升到那一高度所需的时间完全相等。

［5］In each situation it is important to *take a* representative sample by mixing the sand or by taking multiple samples in different locations.

译文:每种情况下,选取混合砂或在不同位置用多个样品作为代表性样品是很重要的。

［6］Take reading every ten minutes.

译文:每 10 分钟*测取*一次读数。

［7］Care has to be taken before closing the apparatus that carbide and sand do not come into contact.

译文:在关闭设备前必须小心,不要让电石和型砂接触。

2）根据专业内容选择词义

由于科学技术的发展,各个专业相互渗透、相互影响,又相互联系,这就造成了词义变通和交叉使用的现象,尤其是专业英语中大量使用半专业词汇,因此在理解和选择词义时,必须考虑文章中所论学科和专业。一旦确定了其在文章中的专业范围,其词义就比较明确了。

下面以"converter"为例,不同的专业领域,有着不同的意义。

［1］As example, analysis automation for a *converter* steelmaking works is introduced.

译文:文中还介绍了一个*转炉*炼钢厂分析检测自动化的实例。（冶金）

［2］This paper firstly introduced the structure of the wind turbine and its—*converter*, and analyzed causes of *converters*' faults.

译文:首先介绍了风力机及其*变频器*系统的结构,分析了*变频器*的故障机理。（物理）

［3］With the development of modern industry, more and more attention has been on the *converter*, such as inverter power and high frequency switch.

译文:随着现代工业的发展,越来越多的目光落在了*整流*装置上,如逆变电源和高频开关等。(电工)

3)根据搭配选择词义

由于英汉遣词造句的方式和习惯不同,因此,一个词的词义在不同的搭配中具有不同的意义。这种搭配的类型有:动词和名词,动词和副词或介词,以及形容词与名词。值得注意的是,一个词与不同的词搭配,在较多的场合里,其语义是有差异的。另外,英语中的习惯搭配和固定搭配,必须根据汉语的搭配习惯来选择适当的词义。

［1］These simulators evolved from an original device *developed* by Nippes and Savage in 1949.

译文:这些模拟仪是在 Nippes 和 Savage *开发*的原始设备上逐渐发展起来的。

［2］From the one metal iron, probable more than 25,000 alloys of steel are found to have been *developed*, although the uses of pure iron are very few in contrast to the vast range of purpose served by the many steel alloys.

译文:虽然同很多合金钢相比,纯铁本身的用途很少,但就纯铁这种金属来说,已经*发展*出 25 000 多种合金了。

［3］To overcome these difficulties transfer molding has been *developed*.

译文:为了解决这些困难,*发明*了传递模塑。

［4］The amount of heat which a fuel will *develop* depends on its calorific value.

译文:某种燃料产生的热量取决于它的发热值。

［5］Reactors of improved design, greater power, and operable at high temperature were rapidly *developed* in U. S.

译文:美国很快*建造*了一些设计更先进、功率更大且在高温下运转的反应堆。

［6］This technology has *developed* high technical proficiency.

译文:这项工艺已*达到*高度的技术熟练程度。

分析:以上六个例句中,develop 跟不同的名词宾语(在被动结构中则是该词与用作主语的名词)搭配,英译汉时采用了符合汉语习惯的不同搭配。

4)根据上下文选择词义

任何一个词,即使是多义词,当它用在一段具体的文字中,其意思却是确定、单一的。至于这个词在汉语中确定含义到底是什么,只能根据该词所处的语言环境,联系上下文,从词的基本含义出发,通过认真思考,体会作者通过该词表达的真正意图,从而选择贴合原意的合适的释义。

［1］If the fluid reacts with active metals such as silver and copper alloys, pump and valve *designs* must change.

译文:如果流体与活泼金属(例如银和铜合金)发生反应,那么必然改变泵和阀的*形状*。

分析:design 原义为设计,pump and valve designs 译成"泵和阀的设计"似乎准确无误,从整个句子中词与词的联系来看,"……发生反应,……必然改变……设计"则说不通。"改变设计"只能由人来进行,而不能由反应来进行,"……发生反应"只能"改变"泵和阀的"形状"。因此,design 在句子中译成"形状",则词义比较准确。

[2] Effective control of the *chip*, when it moves across the face of the tool, is best achieved by guiding the *coil* so that it takes a tightly wound helical or spiral formation.

译文:切屑通过刀具表面时,控制切屑的有效方法是引导切屑呈紧密缠绕的螺旋状。

分析:chip 和 coil 并非同义词:前者为"(金属等的)切屑",后者为"线圈""绕组"。但从上下文看,两者具有词汇同义关系,coil 失去原义,仅用于代替 chip,以避免用词雷同。

从以上两个例句中不难发现,其中不少词的汉译用词是从词典释义中查不到的,但选用这些词是符合作者表达的意图,贴合原义的。

38.2.2　词义的引申

所谓词义引申,是指根据语篇或语境对一个词的字面意义或词典上所提供的释义作一些必要的调整变动。在专业英语汉译过程中,经常会遇到一些词或词组,无法直接照搬词典中的释义译出,如果勉强按照词典所给的释义译出,就会使译文晦涩难懂甚至会引起误解。遇到这种情况,我们就需要从词的基本意义出发,根据上下文的逻辑联系和汉语的搭配习惯等对词义进行必要的调整变动或者说对词义进行引申。之所以要对词义进行引申,是因为词义的复杂性所致。我们知道,一个词不仅有直接的、表面的、一般的意义,还有比喻的、内涵的、特定的意义。词典上所给的意义,勾画出词义的一个粗疏的轮廓,或者说,只是词的最基本的含义,而不是词的全部意义。由此看来,词义的引申与语篇或者语境有着密切的联系,语篇或语境是词义引申的基础。

[1] Electric power *became the servant of* man only after the motor was invented.

译文:只是在电动机发明之后,电力才开始为人类服务。

分析:句中 became the servant of 本义是"成为……的仆人",意义很具体,很形象,比喻"为……服务",但直译不符合汉语的习惯,因此根据其内涵和比喻意义,作抽象化引申为"为人类服务"。

[2] These sections of sand, which protrude into the hole in the pattern, are called cores. In this case, because they are made of *green sand*, they are called green-sand cores.

译文:这些凸出于砂模中的部分的砂称为砂芯。此时,由于它们由湿砂制成,被称为湿砂芯。

分析:英汉翻译中,经常需要将原文中含义比较抽象而概括的词用意思比较明确而具体的词表达出来,以避免译文晦涩费解。在翻译这一类词时,就应该掌握该词的确切含义,从而在译文中作具体化的引申。green sand 在铸造中含义为"湿砂",green 本义为"绿色的、不成熟的",如果翻译成"绿色的砂",就跟作者的原义大相径庭了。

[3] These compounds are mixtures of materials that when heated will generate a gas that will *give up* its carbon to austenite.

译文:这些渗碳剂是几种材料的混合物,当加热时,它会产生一种气体,这种气体可以释放出向奥氏体扩散的活性炭原子。

分析:give up 原义为"放弃",这里将其引申为"释放"。

[4] One of the advantages of *pack carburizing* over other methods lies in the possibility of using almost any type of furnace that will hold the boxes, as long as the furnace is capable of maintaining uniform temperature.

译文:与其他渗碳方法相比,固体渗碳优点在于它可以使用任何装有渗碳箱的炉子,只要炉子能够保持均匀的温度即可。

分析:英语中有时用表示具体形象的词或短语来表示某种特性、事物、概念等。译成汉语时,往往要将这种含义或短语作抽象化的引申,用比较笼统概括的词加以表达,以使译文明快达意。pack 一词的基本意义是"(把……)打包;塞进;拥进;(使)聚集成团",这里根据上下文,以及跟 carburizing (渗碳)搭配在一起,将 pack 的词义引申为"固体"。

[5] A high voltage across the filament(cathode) causes it to emit electrons, which are then *driven* to the target(anode) where the sudden deceleration results in X-rays.

译文:阳极靶上的高电压使其发射出电子,这些电子在高压下运动直接打到阳极靶上并突然停止,从而产生 X 射线。

分析:drive 的基本词义是"使移动;驱动;推动",这里根据语境将其词义引申为"运动"。

值得注意的是,词义引申的目的是为了使译文更忠实、更通顺、更完整地表达原文的意义。因此,引申必须得当适度,切忌忽略原文固有的基本含义、脱离开上下文的逻辑联系而妄加发挥。词义引申的一个重要原则是必须从原文固有的基本含义出发,不脱离上下文的联系,不任意发挥。这样,就能从日常的翻译实践中积累经验,掌握好词义引申这一翻译技巧。

38.2.3 词语的增译

由于英汉两种语言的表达方式不同,词汇不能够一一对应,这就意味着翻译过程中,需要增补一些必要的词语来使译文既通顺,又符合汉语的行文规范和表达习惯。

1)增加名词

专业英语中经常使用一些由形容词或动词派生出来,表示行为动作的抽象名词,作者在使用这类词时,往往表示一些较为具体的概念。为了体现实际意义,使译文明确,翻译时要根据上下文和专业习惯,选择适当的名词添加在他们的后面(构成一定中短语形式的名词词组)。用来在有动作意义的抽象名词后面添加的名词有量、值、机器、设备、装置、系统、作用、现象、概念、方式、方法、途径、效应、要求、结果、事业、能力、情况、状态、过程、进程、研究、技术、试验、问题、信号、机制、形式、任务、工作等。例如:induction 感应现象,anisotropy 各向异性现象,representation 表示法,constraint 约束条件,distribution 分布曲线,electrospaking 电火花加工,tunneling 隧道效应,gearing 传动装置。另外,在一些形容词单独译出表示的意义不明确时,也需要添加名词,以明确其意义。

[1] The *expansion* of metal on heating must be taken into consideration before a long metal bridge is built.

译文:要建造金属大桥,就必须考虑金属的**热膨胀特性**。

[2] Vertical *movement* of arm on column is 520 mm.

译文:摇臂在立柱上的垂直**移动量**是 520 mm。

[3] As compared with other welding process, CO_2 welding is high in efficiency, low in cost and ready for automatization.

译文:与其他的焊接方法相比,CO_2 焊具有高效率、低成本且易于实现自动化的**特点**。

[4] Were there no electric pressure in conductor, the electron *flow* would not take place in it.

译文:导体内如果没有电压,便不会产生电子**流动现象**。

[5] In most cases, small lathes are as complete as large lathes, only *smaller* and *lighter*.

译文:小车床的结构与大车床一样完备,只是**体积小些**,**质量轻些**。

2)增加动词

在专业英语文献中,通常用来添加汉语动词有加、求、定、使、进行、举行、产生、引起、发生、了解、成立、发挥、采用、利用、选择、确定、配备、作出、编制、提高、遭受、陷于等。例如:setting 调整位置,programming 编制程序,loading 加负载,summation 求和等。

[1] The problem of acquiring, tracking and discriminating all the targets needs the full *capability* of the phased array radar.

译文:解决捕获、跟踪和识别所有目标的问题,要求**发挥**相阵控雷达的全部能力。

[2] Electronic *methods* allow for great adventure of space travel is a space station.

译文:**采用**电子控制方法能使控制系统的操作更加迅速、更准确,而且更灵活。

3)增加连贯语气的词

原文有些句子的各个成分之间、句子与句子之间,表达因果、转折及顺理连接等关系时,需要增加连贯的词才能将原文的层次、条理和关系表达清楚。

[1] Manganese is a *hard*, *brittle*, grey-white metal.

译文:锰是一种灰白色,**又硬又脆**的金属。

[2] Carbon combines with oxygen to form carbon oxides.

译文:碳与氧化合后**便**构成碳氧化合物。

[3] Other liquids being too light, a barometer uses mercury.

译文:**由于**其他液体太轻,**因此**,气压表采用水银。

4)增加概括性词

英语中在列举事实时很少运用概括性词语,但汉语则恰恰相反。因此英译汉时应根据情况补充一些概括性词语。

[1] The chief effects of electric currents are the magnetic, heating and chemical effects.

译文:电流的主要效应有磁效应、热效应和化学效应**三种**。

[2] This report summed up the new achievements made in electron tubes, semiconductors and components.

译文:这篇报告总结了电子管、半导体和元件**三方面**的新成就。

5）增加解释性词语

[1] The electronic tube is unstable, power hungry and easy to burn out. It also takes up much space.

译文：电子管不够稳定，耗电量大，容易烧坏，占据的空间也大。

[2] This is what we must handle at first.

译文：这是我们必须首先处理的*情况*。

38. 2. 4　词语的省译

由于英汉两种语言表达习惯不同，翻译时如果一字不漏地照搬译出，往往会显得累赘或不合行文习惯，甚至产生歧义，采取省译这种译法可以使译文言简意赅，表达流畅。

1）冠词的省略

冠词是英语中使用频率最高的功能词之一，而汉语中没有冠词，也没有与之相当的词类，因此在英译汉时，除了非译不可外，经常略去不译。

[1] *The* casting process is extensively used in manufacturing because of its many advantages.

译文：铸造方法由于其许多优点而广泛用于制造业。

分析：the 这里指特定的事物，相当于汉语指示代词"这"和"那"。但在很多情况下，它所指示的意义可以借助于上下文的语言环境衬托出来。凡是遇到这种情况，一般可以省去不译。

[2] Any substance is made up of atoms whether it is *a* solid, *a* liquid, or *a* gas.

译文：任何物质，不管它是固体、液体或气体，都是由原子组成。

[3] As has been defined earlier, *a* pattern is *a* replica of the object to be made by the casting process, with some modifications.

译文：如前所述，模样是经修改过的铸造复制品。

分析：a 指人或事物的类属，在这种情况下，汉译时应略去不译。

2）代词的省略

英语中代词的使用比较频繁，在翻译成汉语时可以将其省略。

[1] By the word "alloy", *we* mean "mixture of metals".

译文：用"合金"这个词来表示"金属的混合物"。

[2] If *you* melt two or more metals together, *you* can get a new metal.

译文：把两种或多种金属熔化在一起，就可获得一种新金属。

[3] Hardened steel is difficult to machine, but having been annealed, *it* can be easily machined.

译文：淬火钢很难加工，但经过退火后便很容易加工。

分析：英语句子中每句必有主语，常在后句中用人称代词表示和前句相同的主语。一段文字中人称代词常常出现多次，这样使用的人称代词在汉译时应省略不译。第二句例句中"you"在专业英语文献中往往是泛指的，具有"概括"或"不定"的意义，这种情况，大多也不需要翻译。

3）名词的省略

（1）介词 of 前面的名词的省略

英译汉时，经常省略介词 of 前面表示度量等意义的名词。通常情况下，of 后面的名词可以完整地表达意思，如果把 of 前面的名词译出来反而显得多余。

[1] The size *of the mould* should be a little larger as the casting shrinks as it cools.

译文：因为铸件冷却时收缩，所以**铸模**应该大一些。

[2] Gonging through *the process of heat treatment*, metals become much stronger and more durable.

译文：金属经**热处理**后，强度更大，更加耐用。

（2）同义词或近义词的省略

专业英文文献中，常出现两个同义词连用的情况，即对同一概念用两个词重复表达。这两个词前一个往往是专业词汇，而后一个往往是普通词汇。两者在语法上是同位关系，常用连词 or 连接。汉译时，只需要译出其中一个，以免文字重复。

[1] The surface grinder is a machine for finishing *flat or plane* surface of a work.

译文：平面磨床是对**工件平面**进行精加工的机床。

[2] The ability to make a freehand *drawing or sketch* is an essential skill for every engineering.

译文：**徒手制图**的能力是每一位工程师所必需的技能。

4）介词的省略

英语使用介词非常频繁，汉语中如果逐词译出，势必用很多虚词显得累赘。因此，在翻译的时候可以将一些介词省略，比如说表示时间、地点的介词以及某些介词短语中的介词等。

[1] The critical temperature is different *for* different kinds of steel.

译文：不同种类的钢，其临界温度也不同。

[2] In hot working, the temperature at which the working is completed is important since any extra heat left after working will aid *in* the grain growth, thus giving poor mechanical properties.

译文：在热加工中，工件停止加工温度是很重要的，因为加工后留下的任何余热将有助于晶粒生长，从而使力学性能变差。

[3] Rolling is a very economical process for producing large volumes of material *with* constant cross-section.

译文：轧制是用于生产定截面的大体积材料的非常经济的工艺。

5）连词的省略

汉语的连接词用得不多，其上下逻辑关系常常是暗含的，由词语的次序来表示。英语则不然，连接词用得比较多。因此，在英译汉时，很多情况下不必把连接词译出来。

[1] Metal as well as no-metal expands when heated.

译文：金属以及非金属受热都会膨胀。

〔2〕If there were no heat-treatment, metal could not be made so hard.

译文:没有热处理,金属就不会变得如此硬。

〔3〕If alloyed with tin, copper forms a series of alloy which are known as bronze.

译文:铜与锡结合,就形成称为青铜的一系列合金。

6)动词的省略

英语中一个句子总少不了一个充当谓语的动词,而汉语中却不一定非用动词,汉语句子可以用名词、形容词等来担任动词的表达。因此,汉译时可根据具体情况把原文的某些动词省略不译。

〔1〕The medium carbon steel *has* a high melting point.

译文:中碳钢熔点高。

〔2〕In conduction and convection energy transfer trough a material medium *is involved*.

译文:在传导和对流时,能量通过某种介质传递。

〔3〕Stainless steels *possess* good harness and high strength.

译文:不锈钢硬度大、强度高。

38.2.5 词性的转换

英语和汉语分属于不同的语系,英汉两种语言在语言结构上有许多差异。因此,在英译汉时就不能因循于英文的表面语言形式,而是要在准确再现原文信息的前提下,将原文中的某些词性灵活地转换为汉语的其他词性,如将英语中的名词、介词、形容词、副词转换成汉语动词,或将英语中的动词、形容词、副词、代词转换成汉语名词等。

1)名词、介词、副词、形容词转化成动词

与英语的表达不同,汉语中的动词用得比较多。英语中只用一个动词作谓语,而汉语中可以几个动词或动词性结构连用。因此,英语中不少词类,尤其是名词、介词、形容词、副词,在翻译时,可以转译成动词。

〔1〕Second, because of the surface heat losses, the temperature at the surface can be *lower* than that at the centerline of the specimen, especially if the peak temperature is high and the thermal conductivity of the specimen is low.

译文:第二,由于表面热损失,表面的温度可能**低于**样品中心线的温度,特别是如果峰值温度高并且样品的热导率低。(形容词转换为动词)

〔2〕By *use* of ultrasonic waves, one can find out if there is a flaw in the metal.

译文:人们使用超声波能够发现金属是否有缺陷。(名词转换成动词)

〔3〕In modern times, industrial forging is done either *with* presses or *with* hammers powered by compressed air, electricity, hydraulics or steam.

译文:在现代,工业锻造是**用**压力机或空气锤、电力锤、液锤压或蒸汽锤。(介词转换成动词)

〔4〕The experiment is *over*.

译文:实验**结束**了。

2）动词、形容词、副词转译成名词

[1] The vertical spindle drilling machine *is characterized* by a single vertical spindle rotating in a fixed position.

译文：立式钻床的**特点**是具有单独一根固定位置上旋转的垂直主轴。（动词转成名词）

[2] The cutting tool must be *strong*, *tough*, *hard*, and wear resistant.

译文：刀具必须有足够的强度、韧性、硬度，而且要耐磨。（形容词转成名词）

[3] The blueprint must be *dimensionally* and *proportionally* correct.

译文：蓝图的**尺寸**和**比例**必须正确。（副词转成名词）

3）名词、副词转译成形容词

[1] In certain cases friction is an absolute *necessity*.

译文：在一定场合下，摩擦是绝对**必要**的。（名词转成形容词）

[2] Gases conduct *best* at low pressures.

译文：气体在低压下导电性**最佳**。（副词转成形容词）

4）形容词、动词转译成副词

[1] A graph gives a *visual* representation of the relationship.

译文：图表可以直观地显示要说明的关系。（形容词转换成副词）

[2] Applying electric - Kanban system to realize pull JIT system and reduce high inventory, reduce the various wastes and non-value added activities caused by materials handling to *succeed* in the integration between material flow, information flow and cash flow.

译文：应用电子看板系统实现拉动式的准时化管理模式达到降低库存，减少物料流通环节的浪费和过多的不增值活动，并**成功地**实现了信息流、物流及现金流的有机结合。（动词转换成副词）

38.2.6 数词的翻译

1）数字的译法

对于专业英语中数目不大的数字可采用直译法，如温度、年代、数量、高度等。

[1] In the Rockwell B test, a steel of ball of 0.062 5 inch diameter is used with a load of 100 kg.

译文：在洛氏 B 试验中，使用负荷为 100 kg，直径 0.062 5 英寸的钢球。

2）不定数量的译法

不定数量是指表示若干、许多、大量、不少、成千上万等概念的词组。英语能够表示这些概念的词组主要有以下几类：

（1）在 number, lot, score, decade, dozen, ten, hundred, thousand, million 等词后加复数后缀 -s，这些词组都有约定俗成的译法。例如：

a hundred and one	许多；无数	dozens of	几十；几打
a thousand and one	许多；无数	lots of	许多；大量

scores of	几十;许多	thousands of	几千
millions of	千千万万	millions upon millions of	数亿;无数
teens of	十几	decades of	数十;几十

[1] The number of known hydrocarbons runs into *tens of thousands*.

译文:已知的碳氢化合物多达**几万种**。

(2)数字前加 above, more than, over, up to 等,可译成"超过""多达"等。

[1] The temperature in the furnace is not always *above* 1,000 ℃.

译文:炉内的温度并不是每次都**超过**1 000 ℃。

[2] Alloys containing *up to* 2% carbon are termed steels and *above* 2% are called cast iron.

译文:含**不大于**2%碳的合金称为钢,**高于**2%的合金称为铸铁。

(3)"as…as"可以表示增加,可译为"多达"。

[1] The software approach can control *as many as* eight drives.

译文:这种软件方法能控制**多达**八台磁盘机。

(4)数字前加 below, less than, under 等,译为"以下""不足"。

[1] The efficiency of the best of these steam locomotives is *under* 16%.

译文:这些蒸汽机中,最佳者的效率也**不足**16%。

[2] It is not possible to get any hardness in low carbon steels(*less than* 0.3% C).

译文:低碳钢(含碳量**小于**0.3%)不可能获得任何硬度。

(5)数字前加 about, around, close to, nearly, or so, some, toward 等,可以译为"左右""将近"等。

[1] Major iron ores are haematite and magnetite, which contain *about* 55% iron.

译文:主要铁矿石是赤铁矿和磁铁矿,其含有**约**55%的铁。

(6)from…to, between…and 等可译为"从……到""到"等。

[1] Crystals of austenite are separated from the liquid along the line BC with the composition ranging *from* 0.18% *to* 2.0%。

译文:奥氏体晶体沿着线 BC 从液体中析出,化学成分范围为 0.18%**至**2.0%。

[2] Though cast irons can have a carbon percentage *between* 2 to 6.67, the practical limit is normally *between* 2% and 4%。

译文:虽然铸铁含碳量在2% ~6.67%但通常实际控制在 2% ~4%**之间**。

3)分数的表示法

(1)公式法

英语中分数结构:分子(用基数词)/分母(用序数词,分子超过 1 要用复数)。如:one-third 三分之一,three-fourths 五分之三等。

[1] This is only *a few thousands of* the heat of vaporization.

译文:这仅是汽化热的**千分之几**。

（2）用 part 表示

［1］Bronze was made by blending copper with about one *part in ten* of tin.

译文：青铜时有紫铜和锡约按**十分之一**的比例混合制成的。

（3）分数+the（what/that）结构

［1］The Fahrenheit degree is only 5/9 the size of the Celsius degree.

译文：华氏度仅为摄氏度数值的九分之五。

4）倍数的译法

（1）几倍于的译法

①倍数+形容词/副词比较级+than

英语中表示"增加了……倍"时，是连原来的基数包括在内的，并不表示纯增加倍数。所以，汉译成"增加了……倍"时，应将英语的倍数减一。

［1］This machine is *four times* heavier than that one.

译文：这台机器比那台**重三倍**。

②倍数+as+形容词/副词+as

"*n* times as…as"的结构可译成"是……的 *n* 倍"或"为……*n* 倍"。

［1］The thermal conductivity of metals is *as much as several hundred times* that of glass.

译文：金属的热导率**是玻璃的数百倍**。

③倍数+that（what）

英语中"*n* times+名词（that/what）"的结构可译成"是……的 *n* 倍"。

［1］For a narrow and deep runner，the well diameter should be 2.5 times the width of the runner in two-runner system，and twice its width in one-runner system.

译文：对于深而窄的横浇道，在双横浇道系统中，（直浇）道窝的直径是横浇道宽的 2.5 倍，在单横浇道系统中，（直浇）道窝的直径是横浇道的宽的两倍。

［2］The demand for this kind of equipment in the near future will be 20 *times what it is*.

译文：不久的将来，对这种设备的需求量将为目前的 20 倍。

［3］The resistivity of aluminum is 1.6 *times* that of copper.

译文：铝的电阻率是铜的*1.6 倍*。

（2）倍数和数量增加的译法

①表示增加意义的动词+倍数

increase+…times+名词（或代词），可译成："为……的 *n* 倍"，或"增加了 *n*−1 倍"；increase+by+…times 或百分比，可译成：增加了，如增加了 50%，指原来为 100，现在为 150。

［1］Sodium bentonites produce better swelling properties—volume *increases some* 10 *to* 20 *times*，high dry strength and high resistance to burnout，which reduces clay consumptions.

译文：钠基膨润土产生更好的溶胀性能，体积*增加 10 至 20 倍*，由于高干强度和高耐燃烧性，这减少了黏土消耗。

［2］During the last six years the steel output in the works has *multiplied by* 85%.

译文：在近 6 年中，这个厂的钢产量*增加了 85%*。

［3］In this case both pulse spacing and the group length are *increased by* 8 *times*.

译文:在这种情况下,脉冲间隔及脉冲组长度*增加了7倍*。

②表示倍数意义的动词+宾语

专业英语中表示倍数意义的动词主要有 double(增加 1 倍),treble(增加 2 倍),quadruple(增加 3 倍)等。

［1］If the resistance is *doubled* without changing the voltage, the current becomes only half as strong.

译文:如果电压不变,电阻*增加 1 倍*,电流就减少1/2。

［2］The efficiency of the machines has been more than *trebled or quadrupled*.

译文:这些机器的效率已经提高了*2 倍或 3 倍多*。

(3)倍数或数量减少的译法

英语中使用"减少意义的词+倍数 n"来表示减少的数量,翻译时要译成"减少了(n-1)/n"(指减去部分),或译成"减少到1/n"(指剩下部分)。

［1］By using this new process the loss of metal *was reduced four times*.

译文:采用这种新工艺使金属损耗量*减少了3/4*。

［2］The hydrogen atom is nearly *16 times* as light as the oxygen atom.

译文:氢原子的重量约为氧原子的*1/16*(即比氧原子约轻 15/16)。

38. 2. 7 专业术语的翻译

专业术语是指自然科学和社会科学领域里的专业性名词。由于现代新技术的迅速发展,在专业英语文献中常常会出现一些新词。翻译时必须首先弄清原词的科学技术含义,再选择或创造相应的汉语术语,以保证翻译的准确性。常用的翻译方法有意译法、音译法、象译法和形译法。

1)意译法

意译,即根据原文的含义译为相应的汉语。这是专业术语翻译最基本的方法。专业术语应尽量采用意译,以便于读者能直接理解新术语的确切含义。例如:

monocrystal	单晶	skin effect	集肤效应
anisotropic	各向异性	grinding tools	磨削工具
nonmetal	非金属的	ferromagnetic	铁磁的
postheat	后热	transfer molding	递模法
cermet	金属陶瓷	overload	过载

2)音译法

有些专业术语有时采用意译不方便,可采用整体音译法或部分音译法。音译法常用于新材料、新发现,以及计量单位等。在一些以人名命名的专业术语中,还可以采用意、音混合译法。例如:

invar	因瓦合金	permalloy	坡莫合金

austenite	奥氏体	Boltzmann's constant	波尔兹曼常数
mullite	莫来石	Hall effect	霍尔效应
nanometer	纳米	nylon	尼龙
weber	韦伯	ohm	欧

3）象译法

在专业英语文献中，常用字母或词描述某种事物的外形，汉译时也可采用同样方法，用具体形象来表达原义，称之为"象译"。例如：

T-square	丁字尺	cross-riveting	十字铆接
I-steel	工字钢	C-washer	C 形垫圈
U-nut	U 形螺母	zigzag dislocation	Z 字形位错

4）形译法

在专业文献中当遇到某些代表某种特定概念的字母，以及涉及型号、牌号、商标、元素名称时可以不必译出，而直接照抄即可，称之为"形译"。例如：

X-ray	X 射线	Z-cut	Z 切割
CNC	数控机床	CCT	连续冷却转变
V-process	V 法铸造	P-N-P junction	P-N-P 结

38.3　专业英语句子的翻译

专业英语句子相对比较复杂，翻译过程中有必要针对科技英语句子的特点，掌握一些翻译方法，灵活运用一些必要的技巧，以保证忠实原文，并符合科技文献简练、通顺的特点。英语句子结构和汉语句子结构有着较大的差异，句子处理的好坏直接影响着翻译的质量，是顺利完成专业英语文献翻译的基础。

38.3.1　简单句的翻译

英语中的简单句无论在形式上还是在内容上都相对复句要简单得多，绝大多数的简单句可按原文结构形式进行直译。但是我们经常碰到一些看似简单的句子，汉译后不太符合汉语的表达习惯。这主要是因为英汉两种语言表达方式不同，句子结构不同，所以翻译时需要转换句子的成分，才能使译文通顺易懂。

1）句子成分转换

（1）主语的转译

[1] *Machinery* has made the products of manufactories very much cheaper than formerly.

译文：因为**机械化**的缘故，工厂里出的产品比起以前来，价格便宜多了。（转译成状语）

(2)非主语译成主语

［1］The electric arc may grow to an inch in *length*.

译文:电弧**长度**可以增长到一英寸。(介词宾语译成主语)

［2］Conductors have very small *resistances*, and the smaller the resistance, the better the conductor.

译文:导体的**电阻**很小,电阻越小,导体越好。(动词宾语译成主语)

［3］Two widely used alloys of copper are *brass* and *bronze*.

译文:**黄铜**和**青铜**是两种广泛使用的铜合金。(表语译成主语)

(3)非谓语译成谓语

［1］Manganese has the same *effect* on the strength of steel as silicon.

译文:锰对钢的强度的**影响**和硅相同。(定语译成谓语)

［2］Precautions are necessary to prevent it from burning.

译文:必须**注意**不要让它烧着。(主语译成谓语)

［3］This explanation is *against* the natural laws.

译文:这种解释**违反**自然规律。(介词译成谓语)

(4)非宾语译成宾语

［1］*The mechanical energy* can be changed back into electrical energy by means of generator.

译文:利用发电机可以把机械能再转变成**电能**。(主语译成宾语)

［2］Various substances *differ* widely in their magnetic characteristics.

译文:各种材料的磁特性有很大的**不同**。(谓语译成宾语)

(5)非定语译成宾语

［1］*Medium carbon steel* is much stronger than low carbon steel.

译文:**中碳钢**的强度比低碳钢大得多。(主语译成定语)

2)语序的调整

语序指各级语言单位在组合中的排列次序,是语言的重要组合手段之一,既反映了一定的逻辑事理,又体现了人们在长期使用语言过程中所形成的语言习惯,还反映了语言使用者的思维模式。中英文描述同一客观事实有不同的语言表达顺序,翻译时有必要作适当调整。

(1)定语语序调整

英语的定语语序与汉语的差别较大,英语的定语既可以前置又可以后置(称为前置定语和后置定语),而汉语中定语只可以前置,因此翻译时要调整语序。

［1］This is *quite a different* situation from static loading, especially in ductile metals.

译文:这跟静载荷作用的情形**十分不同**,特别是在韧性金属中。

［2］Metals can be made into any shape *desired*.

译文:金属可以制成任一**需要**的形状。

［3］Substances having very *high resistance* are called insulators.

译文:电阻**非常高**的物质称为绝缘体。

（2）状语语序的调整

英语中状语的位置较为灵活，可以位于句首或句末，也可以位于主谓语之间或谓语的两个部分之间，修饰动词的状语和整个句子发生关系的时间和地点状语更是如此。但是，在汉语中，状语的位置却是固定的，大多位于主谓语之间，有时可位于句首，很少位于句末。

［1］Being alloyed with certain metals, aluminum can be strengthened.

译文：铝和某些金属熔合后，强度会增加。

［2］Pressure transducers respond more *rapidly* to pressure changes.

译文：压力传感器对压力的反应**迅速**得多。

［3］We shall deal with the behavior of a PN junction *later*.

译文：我们将**在后面**讲到 PN 结的特性。

38.3.2　长句的翻译

对于复杂长句，要抓住句子结构，即辨清主谓（宾），才能把握全句。英语具有科学的语法形态，句子结构的成分分析时翻译时显得尤为重要。简单的句法辨析容易做到，然而长句比较复杂，在翻译时，英语的长句结构容易使以汉语为母语的人产生思维混乱，难以一下把握其真正意义。复杂长句的处理方式一般采用顺译法、逆序译法、调序翻译法和句意反译法。

1）顺译法

如果英语句子依照时间先后顺序来描述事件的发生过程，是符合汉语句子的表述方式的。在英汉语言转换过程中，一般可按照原句的顺序翻译。

［1］The art of mixing metals was gradually developed and it became known that "alloy" formed in this way was sometimes stronger, harder and tougher than the metals of which it is composed.

译文：把金属混合起来的工艺是逐渐发展起来的，人们还知道了用这种方法构成的合金有时比组成这种合金的金属更坚固，更坚韧。

［2］There are many factors influencing hot die forging quality of turbine blade, such as material property, billet outline and size, die forging method and plan, die cavity structure and drop forging equipment, etc.

译文：影响涡轮叶片热模锻质量的因素有很多，例如材料性能，钢坯轮廓和尺寸，模锻方法和工艺，模腔结构和锻压设备等。

［3］There is evidence that early workers understood that higher contents of tin were used the alloy was harder, while less tin gave a softer alloy, so that for different purposes bronzes with varying tin contents were deliberately produced.

译文：有证据表明，早期的工人懂得，锡的含量越大，合金越坚硬，含量越小，合金越软。因此，可以根据不同的用途精心制造出不同含锡量的青铜。

2）逆译法

当英语长句的顺序与汉语表达方式不一致时，常常使用转换、颠倒、改变部分或完全改变词序的逆译法。

[1] *The final part of the paper* highlights the areas for potential application of thermomechanical treatments and emphasizes the *need* for information to facilitate design of suitable forming equipment for exploiting the potential of thermo-mechanical treatments over a range of temperatures as a means of producing various product forms with enhanced property combinations.

译文:文章最后部分着重论述了形变热处理可能应用的领域,并强调指出:为了便于设计适宜的成型设备,作为生产具有优异综合性能的不同形状产品的工具、需要获得有关的知识,而这种成型设备可用来发挥在一系列温度下形变热处理的潜力。

分析:该长句是含有一个主语、两个动词的简单句;因第二个动词的宾语 need 带有很长的定语,所以将它分出来另译成一个句子形式(译文冒号后面由两个并列分句组成的宾语从句)并且打乱了原来的词序,部分地采用了逆译法。

3)拆译法

英语的句子长,动词少,语序灵活,若完全按照原文的形式译出,会使译文结构臃肿,逻辑不清。在这种情况下,可以按照原文的主次关系和逻辑层次把其中某一部分分译出来,化整为零,可以使原文表达更为准确。

[1] Steel is usually made where the iron ore is smelted, so that the modern steelworks forms a complete unity, taking in raw materials and producing all types of cast iron and steel, both for sending to other works for further treatment, and as finished products such as joists and other consumers.

译文:通常炼铁的地方也炼钢。因此,现代炼钢厂是一个配套的整体,从运进原料到生产各种类型的铸铁与钢材;有的送往其他工厂进一步加工处理,有的就制成成品,如工字钢及其他一些型材。

分析:通过分析句子结构,可以看出原句共有三个小句:steel is …smelted; so that …steel 和 both for …consumers。通过 both …and 连接的两个介词短语可在译文中扩展成句子。可以将三个小句译为对应的三个独立的汉语小句,这样译文既简洁,又能表达清楚概念。

[2] Nanotechnology has been described as a key manufacturing technology of the 21st century, which will be able to manufacture almost any chemically stable structure at low cost.

译文:纳米技术被认为是 21 世纪一门重要的制造技术,它能够低成本地制造出几乎各种化学成分稳定的结构。

分析:如果非限制性定语从句补充说明某个词时,汉译时常常可以译为并列分句,分句主语可重复先行词,也可用"他(它)""他(它)们""该""这"等词来代替先行词。

4)综合法

对于长而复杂的句子,如果只是采用顺序法、逆序法或拆译法,都会不可避免地使译文结构失调、层次混乱,从而导致理解上的错误。在这种情况下,有必要采用综合法,把原文的结构顺序全盘打乱,按其时间先后、逻辑层次和主次关系重新排列。通过这样的处理,译文会显得脉络分明,其表意更加清楚,不会造成误解。

[1] Radial bearings which carry a load acting at right angles to the shaft axis, and thrust

bearings which take a load acting parallel to the direction of shaft axis are two main bearings used in modern machines.

译文:径向轴承能够承受与轴垂直的载荷,推力轴承主要承受沿着轴向方向的载荷,它们是现代机器上使用的两种主要轴承。

分析:本句有两个并列主语 radial bearings 和 thrust bearings,两个主语后面分别带有一个由关系代词 which 引导的限定性定语从句,其中又分别带有一个由现在分词短语 acting 担任的后置定语。可以先把说明的部分译出来,就是把两个带有定语从句的主语,分别译成两个分句,先作交代,然后再译出句子的主要部分。

38.3.3 特殊句型的翻译

1) 被动语态的翻译

被动语态在专业英语文献中频繁使用。被动语态一般由"be+p. p."构成,介词"by"同动作执行者连用,引出做某事的人或东西,或者介绍做某事的方法,"with"引出做某事所用的工具或器具。汉译时,需要对英语被动句进行转换。常用译法有:

(1) 译成汉语被动句

具有典型意义的英语被动句,即特别强调被动动作或特别突出被动者的被动句,可以保留被动,译成汉语的被动句,通常采用以下几种方法:

①译成谓语前加"被"形式;

②译成谓语前省略"被"形式;

③译成由"被""由""让""受"等引出主体形式;

④译成由"受到""得到""遭到"等形式;

⑤译成"是……的"形式;

⑥译成由"加以""予以"形式。

[1] A large number of observations of pressure, temperature, etc., *are received* and recorded, at the appreciate places, on a large chart.

译文:大量有关气压、气温的观测结果**被接受**并标示在一张大幅图表的适当位置。

[2] Metals *may be cast* into various shapes.

译文:金属**可铸成**各种形状。(省略"被")

[3] All matter *is made up* of atoms.

译文:一切物质都是**由**原子**组成**。

[4] Besides voltage, resistance and capacitance, alternating current *is also influenced* by inductance.

译文:除了电压、电阻和电容外,交流电还**受到**电感的**影响**。

[5] Many casting defects *are caused* by expansion properties of sand.

译文:很多铸造缺陷**是**由砂子的膨胀性质所**造成的**。

[6] Other advantages of our invention will *be discussed* in the following.

译文:本发明的其他优点将在下文中**予以讨论**。

（2）转换成汉语主动句

①译成无主语的动宾关系。

[1] An almost identical cracking problem was encountered in the production of cast 713 turbine blades.

译文:在生产713铸造透平叶片时,碰到了一个几乎相同的裂纹问题。

[2] *Care is to be taken* to remove all the impurities.

译文:**要注意**除去所有杂质。

②当英语被动句中的主语为无生命的名词,且句中一般没有介词引导的行为主体时,这种句子常常可译成汉语的主动句。

[1] When heat is applied to metal, the metal stretches. This action *is called* expansion.

译文:当热施加到金属上时,它就伸长。这种作用**称作**膨胀。

[2] *This method is widely used* for cooling and heating.

译文:**这一方法**被广泛用来冷却和加热。

③句中的介词短语作状语,有时可译成主语。

[1] Gas, oil and electric furnaces *are commonly used for heat treating metals*.

译文:**金属热处理**常用煤气炉、油炉和电炉。

[2] Casting quality *is influenced by a number of factors* such as the nature of metal or alloy cast, properties of mould materials used and the casting process.

译文:**许多因素**能影响逐渐铸件的质量,例如,铸造所用金属或合金,所使用的造型材料和铸造方法等。

④加"人们""我们""有人"等作主语。

[1] Steel and its alloys will still *be taken* as the leading materials in industry for a long time to come.

译文:在今后很长一段时间内,**我们**仍将钢及其合金作为主要工业材料。

专业英语文献中常用的固定被动句式及其翻译有:

be converted into…	把……转换为……
be turned into…	把……转变为……
be separated into…	把……分成……
be grouped into…	把……分成……组
be classified into…	把……分成……类
be categorized into…	把……分成……类
be reduced to…	把……简化为……
be compared to…	把……比喻为……
be built into…	把……造成……
be linked to…	把……联系起来
be shortened to…	把……缩短到……

2)**否定句的译法**

英语否定结构较为复杂,有一系列表示否定概念的语言规则,无论是在用词、语法,还是在逻辑关系上,英语和汉语的表达习惯都有很大差异,因此,在理解与翻译时要特别加以注意。

(1)否定成分的转译

对于一些一般形式上的否定(谓语否定或状语否定),而意义上是特指否定(其他否定),如当否定词位于主语之前时,在意思上否定的往往是谓语,翻译时需要根据汉语习惯将主语否定转译成否定谓语。

[1] There is no perfect conductor and no perfect insulator.

译文:既没有理想的导体,也没有理想的绝缘体。(由否定主语转译否定谓语)

[2] Liquids are different from solids in that liquids have *no definite shape*.

译文:液体与固体的区别在于,液体**没有**一定的**形状**。(否定宾语转译成否定谓语)

[3] Most plastics do not conduct heat or electricity *readily*.

译文:大多数塑料都**不易**传热和导电。(否定状语转译成否定谓语)

(2)部分否定的译法

当由 both, all, every, always 等词与 not 连用时,通常表示"部分否定",译成"并非都……"。

[1] All metals do not conduct electricity equally well.

译文:不是所有的金属都有同样的导电性能。

[2] Both instruments here are not good.

译文:这里的两台仪器并非都是好的。

(3)全部否定的译法

英语中使用 neither, none, no, nobody, never, neither…nor…, nowhere 等词表达全部否定或完全否定。

[1] None of the inert gases will combine with other substances to form compounds.

译文:惰性气体均不会和其他物质化合成化合物。

[2] Neither of the transistors are good.

译文:这两只晶体管没有一只是好的。

(4)双重否定的译法

双重否定通常由 no, not, never, nothing 等词与含有否定意义的词连用构成,汉译时可译为肯定的语气,或根据上下文译成双重否定。

[1] The atomic furnace will *not work unless* it has enough fuel.

译文:原子反应堆若**没有**足够的燃料是**不能**运转的。

[2] No current will flow at any time unless there is a complete circuit.

译文:除非有完整的电路,否则任何时候都不会有电流流过。

(5)其他意义上的否定

专业英语中一些词或词组也表示否定的意思,翻译时应当注意这些表达。例如:

little	几乎没有	scarcely	几乎不……

rather than	而不	fail(加动词不定式)	不能
hardly	几乎不……	by no means	绝不
no amount of	怎么……也不	in no way	绝不
no longer	不再	instead of	而不(代替)
too…to…	太……不	free from	没有

［1］With lens we can see objects *too* small for our eyes *to* examine or *too* far away *to* be seen clearly.

译文:借助于透镜我们能看到那些肉眼察觉不到的微小物体,或太远而看不清楚的物体。

［2］Carbon is no more a metal than sulfur is.

译文:碳和硫一样都不是金属。

3)强调句的译法

英语中强调句子成分的方法有很多,主要有以下几种:

(1)It is(was)+被强调的部分
- 主语
- 宾语
- 状语
  - 副词
  - 介词短语+that(which, who)
  - 状语从句
- 介词宾语(后可跟"介词+which")

这种 It 引导的强调句结构可译成"正是……""就是……""是……"等。

［1］*It is* the ability of the mounding material to withstand the high temperature of the molten metal so that it does not cause fusion.

译文:**正是**这种造型材料能承受熔融金属的高温,从而不会引起熔化的能力。(强调主语)

［2］*It is the losses caused by friction* that we must try to overcome by various means.

译文:我们必须想各种办法来克服的正是由**摩擦引起的损失**。(强调宾语)

［3］It was *from a study of algebraic equations* that mathematician was led to predict that only 32 types of crystals would be found in mineralogy.

译文:**正是**根据代数方程的研究数学家们才能预知在矿物学中只能找到 32 种晶型。(强调状语)

(2)"do+动词"形式

在英语中,借助助动词 do+谓语动词,可以表达对句子谓语部分的强调。这类句式通常可以译成"是""确实""实际""真的""一定""的确"等。

［1］But if the positive charges *did move*, they would flow through the wire from the positive terminal to the negative one, that is, just as it was supposed one hundred years ago.

译文:但如果正电荷**确实运动**的话,它们就从正端通过导线流向负端,正与百年所假定的情况一样。

(3)very 强调名词

形容词 very 强调句子中的某个名词时,译成"正是""就是""最""那个"。

［1］The alternating current is the *very* current that makes radio possible.

译文:交流电*就是*使无线电成为可能的那种电流。

［2］By their *very* nature, FETs are isolated from each other.

译文:*正是*由于场效应管的固有特性,它们是彼此隔离的。

(4)what 从句强调

［1］*What is important* is the fact that like charges repel and unlike charges attract.

译文:重要的是同种电荷相互排斥,异种电荷相互吸引这一点。

参考文献

［1］王快社,刘环,张郑. 材料加工工程科技英语［M］.北京:冶金工业出版社,2013.

［2］ROBERT H. T, DELL K A, LEO A. Manufacturing Processes Reference Guide［M］. South Norwalk: Industrial Press Inc., 1994.

［3］RAO T V. Metal Casting: Principles and Practice［M］. New Delhi: New Age International, 2003.

［4］KALPAKJIAN S, SCHMID S, Manufacturing Engineering and Technology［M］. 5th ed. Beijing: Tsinghua University Press, 2006.

［5］VISWANATHAN S, APELIAN D, DASGUPTA R, et al. ASM Handbook: Casting［M］. 10th ed. ASM International, 2008.

［6］SIAS F R. Lost-wax Casting: Old, New, and Inexpensive Methods［M］. South Carolina: Woodsmere Press, 2006.

［7］DAVIS J R. Tool Materials［M］. Phenix: ASM International, 1995.

［8］ANDRESEN B. Die Casting Engineering［M］. New York: Marcel Dekker, 2005.

［9］XU Q Y, FENG W M, LIU B C. 3D stochastic modeling of grain structure for aluminum alloy casting［J］. Journal of Materials science & Technelogy, 2003, 19(5): 391-394.

［10］CHEN S D, CHEN J C. Simulation of microstructures in solidification of aluminum twin-roll casting［J］. Transactions of Nonferrous Metals Society of China, 2012, 22(6): 1452-1456.

［11］GOPALAN R, PRABLU N K. Oxide bifilms in Aluminium alloy castings - a review［J］. Materials Science and Technology, 2012, 27(12): 1757-1769.

［12］SCHLESINGER M E, KING M J, SOLE K C, DAVENPORT W G. Extractive Metallurgy of Copper［M］. Amsterdam: Elsevier Ltd, 2011: 166, 239, 256-247, 404-408.

［13］SMITH E H. Mechanical Engineer's Reference Book［M］. 12th ed. Amsterdam: Elsevier Ltd, 1998.

［14］DOEGE E, BEHRENS B A. Handbuch Umformtechnik: Grundlagen, Technologien,

Maschinen（in German）［M］. 2nd ed. Berlin：Springer Verlag，2010.

［15］FRIEDRICH O. Anwendungstechnologie Aluminium（in German）［M］. 3rd ed. Berlin：Springer Verlag，2014.

［16］SAKAGUCHIM，KOBAYASHI S. Effect of drawing condition on mechanical properties and molecular orientation of self-reinforced poly（lactic acid）screws［J］，Advanced Composite Materials，2015（24）：91-103.

［17］CUBBERLY W H，BAKERJIAN R. Tool and manufacturing engineers handbook［M］. 4th ed. Michigan：Society of Manufacturing Engineers，1989.

［18］OBERG E，MCCAULEY C J. Machinery's Handbook［M］. 29th ed. New York：Industrial Press，2012.

［19］杨元刚. 英汉词语文化语义对比研究［D］. 武汉：华东师范大学，2005.

［20］张良军，王庆华，王蕾. 实用英汉语言对比教程［M］. 哈尔滨：黑龙江人民出版社，2006.

［21］连淑能. 英汉对比研究［M］. 增订本. 北京：高等教育出版社，2010.

［22］邓云华，秦裕祥，唐燕玲. 英汉句法对比研究［M］. 长沙：湖南师范大学出版社，2005.

［23］韩礼德，等. 科学语言［M］. 张克定，等，译. 北京：北京大学出版社，2015.

［24］关丽，王涛. 英汉语言对比与互译指南［M］. 哈尔滨：东北林业大学出版社，2008.

［25］范荣. 从中西思维差异浅析科技英语长句的翻译［J］. 吉林农业科技学院学报，2015，24（2）：101-103.

［26］杨德霞. 英汉词汇语义对比与翻译［J］. 山西师大学报（社会科学版），2006（S1）：154-155.

［27］何善芬. 英汉语言对比研究［M］. 上海：上海外语教育出版社，2002.

［28］李建军，盛卓立. 英汉语言对比与翻译［M］. 武汉：武汉大学出版社，2014.

［29］蔡基刚. 英汉词汇对比研究［M］. 上海：复旦大学出版社，2008.

［30］王卫平，潘丽蓉. 英语科技文献的语言特点与翻译［M］. 上海：上海交通大学出版社，2009.

［31］秦荻辉. 科技英语语法［M］. 北京：外语教学与研究出版社，2007.

［32］陈莉萍. 专门用途英语研究［M］. 上海：复旦大学出版社，2000.

［33］单胜江. 专门用途英语教学研究：理论与实践［M］. 杭州：浙江大学出版社，2012.

［34］宋晓岚. 专业英语语言特点研究［J］. 湖南大学学报（社会科学版），2001（S1）：159-163.

［35］刘革. 科技英语语篇中的词汇衔接与翻译：以 Genetically Modified Food 的汉译为例［D］. 长沙：湖南师范大学，2013.

［36］胡壮麟. 语篇的衔接与连贯［M］. 上海：上海外语教育出版社，1994.

［37］CALLISTER W D，RETHWISCH D G. Materials Science and Engineering［M］. 8th ed. Hoboken：John Wiley & Sons，2010.

［38］KURZ W，FISHER D J. Fundamentals of Solidification［M］. Switzerland：Trans Tech Publications，1998.

［39］彭宣维.英汉语篇综合对比［M］.上海：上海外语教育出版社，2000.

［40］周晓梅.翻译研究中的意向性问题［J］.解放军外国语学院学报，2007（1）：74-79.

［41］陈晓玲.中国对近代西方科学书籍的翻译［J］.河南教育学院学报（哲学社会科学版），2000（3）：104-106.

［42］方梦之.近半世纪我国科技翻译研究的回顾与评述［J］.上海科技翻译，2002（3）：1-4.

［43］杨寿康.论科技英语的美感及其在翻译中的体现［J］.上海科技翻译，2004（3）：15-18.

［44］刘宓庆.新编当代翻译理论［M］.北京：中国对外翻译出版公司，2005.

［45］刘军平.西方翻译理论通史［M］.武汉：武汉大学出版社，2009.

［46］黎难秋.中国科学翻译史料［M］.合肥：中国科学技术大学出版社，1996.

［47］黄忠廉.李亚舒.科学翻译学［M］.北京：中国对外翻译出版公司，2007.

［48］曾莲英.从语篇视角看科技英语汉译的词义选择［J］.齐齐哈尔大学学报（哲学社会科学版），2010（3）：70-72.

［49］冯伟年.论英汉翻译中词性和句子成分的转换［J］.西北大学学报（哲学社会科学版），2002（4）：182-186.

［50］余高峰.科技英语长句翻译技巧探析［J］.中国科技翻译，2012，25（3）：1-3.

［51］严俊仁.新英汉科技翻译［M］.北京：国防工业出版社，2010.

［52］武力，赵拴科.科技英汉与汉英翻译教程［M］.西安：西北工业大学出版社，2000.

［53］RAO P N.制造技术：第1卷　铸造、成形和焊接［M］.北京：机械工业出版社，2010.

［54］CALLISTER W D.材料科学与工程基础（第5版）（英文影印版）［M］.北京：化学工业出版社，2004.

［55］张军.材料专业英语译写教程［M］.北京：机械工业出版社，2001.